39 Springer Series in Solid-State Sciences

Edited by Peter Fulde

Springer Series in Solid-State Sciences

Editors: M. Cardona P. Fulde H.-J. Queisser

Volume 40 **Semiconductor Physics** - An Introduction By K. Seeger

Volume 41 **The LMTO Method** By H.L. Skriver

Volume 42 **Crystal Optics with Spatial Dispersion and the Theory of Excitations**
By V.M. Agranovich and V.L. Ginzburg

Volume 43 **Resonant Nonlinear-Interactions of Light with Matter**
By A.E. Kaplan, V.S. Butylkin, Yu.G. Khronopulo, and E.I. Yakubovich

Volumes 1 - 39 are listed on the back inside cover

Anderson Localization

Proceedings of the
Fourth Taniguchi International Symposium,
Sanda-shi, Japan, November 3–8, 1981

Editors: Y. Nagaoka and H. Fukuyama

With 98 Figures

Springer-Verlag Berlin Heidelberg New York 1982

Professor *Yosuke Nagaoka,* Ph.D.

Research Institute for Fundamental Physics
Yukawa Hall, Kyoto University, Kyoto 606, Japan

Professor *Hidetoshi Fukuyama,* Ph.D.

Institute for Solid-State Physics
University of Tokyo, Minato-ku, Tokyo 106, Japan

Series Editors:

Professor Dr. Manuel Cardona
Professor Dr. Peter Fulde
Professor Dr. Hans-Joachim Queisser

Max-Planck-Institut für Festkörperforschung, Heisenbergstrasse 1
D-7000 Stuttgart 80, Fed. Rep. of Germany

ISBN-13:978-3-642-81843-1 e-ISBN-13:978-3-642-81841-7
DOI: 10.1007/978-3-642-81841-7

Softcover reprint of the hardcover 1st edition 1982

2153/3130-543210

Preface

This volume contains the proceedings of the Fourth Taniguchi International Symposium on the Theory of Condensed Matter, which was held at Senkari Seminar House of Kwansei Gakuin University in Sanda-shi, Japan, during the period of 3-8 November 1981. The topic of the symposium was "Anderson localization," one of the most fundamental problems in condensed-matter physics.

Since Anderson's classic paper was published in 1958, much theoretical and experimental effort has been performed to study the problem of electron localization in a random potential. Quite recently, Abrahams, Anderson, Licciardello, and Ramakrishnan proposed a scaling theory of the Anderson localization which made it possible to perform microscopic investigations. Rapid progress has followed and we are now getting a coherent picture of the behavior of electrons in disordered systems.

When we organized the symposium, we asked Dr. Anderson to participate in it and to give a review talk on theoretical aspects of the problem. Though he kindly accepted our invitation, he could not come due to a sudden illness. A review talk was given by Professor Thouless who kindly accepted our request to take the place of Dr. Anderson. Fortunately, Dr. Anderson has since recovered from his illness.

This volume consists of three parts. In the first paper of Part I, which is the review paper mentioned above, Professor Thouless discusses various aspects of the theory of Anderson localization with emphasis on the behavior of noninteracting electrons in a random potential. The following seven papers treat in some detail field-theoretical approaches, diagrammatical approaches, and numerical investigations. Then the effects of interaction between electrons are discussed from various points of view. Theoretical results are compared with experiments on doped semiconductors, for example.

Part II is devoted to a newly developing field which is closely related to the problem of the Anderson localization, i.e., the quantized Hall effect.

Though the problem of spin glasses looks different from that of the Anderson localization, they are closely related as phase transitions in disordered systems. Therefore we invited Professor Toulouse to give a review talk on it. His paper is in Part III.

Finally, we would like to mention the activities of the Taniguchi Foundation. Mr. Toyosaburo Taniguchi, who is the former President of Toyobo Co. Ltd. has supported many academic activities in the natural and human sciences for a long time. He conceived an international symposium where a limited number of active researchers are invited to take part in intense discussions. The Taniguchi International Symposium now consists of thirteen divisions including the theory of condensed matter, mathematics, biological physics, brain science, philosophy, and ethnology. In order to understand the philosophy of Mr. Taniguchi, his speech delivered at the reception is given in the Foreword.

On behalf of all the participants, we would like to express our hearty thanks to Mr. Taniguchi and the Taniguchi Foundation.

April 1982 *Yosuke Nagaoka, Hidetoshi Fukuyama*

Preface

[illegible]

April 1982 Yosuke Nagaoka [illegible]

Foreword

Speech by Toyosaburo Taniguchi

My Dear Friends:

I am very pleased and honored to be here this evening with so many distinguished guests and scholars.

I wish to extend a warm and sincere welcome to those who have come to attend the Fourth International Symposium on the Theory of Condensed Matter, of the Taniguchi Foundation, from within this country and from all over the world.

I am convinced that in order to effectively translate the objectives of the Foundation into action with the funds available, it is necessary to select those fields which are not necessarily in the limelight of popular interest, which means those fields which, I am afraid, are low in funding.

I would rather choose for the foundation projects from modest academic fields than those that stand out in gaudy, gorgeous popular acclaim.

I would desire to see the promising younger groups of scholars assume an even more active role in our symposia, for it is really the younger generation who will lead us into the promise of tomorrow.

I would like to add that we would desire to limit the number of the participants to close to 15, because of my desire to see a situation of living and studying together in a close-knit community setup. It is my hope that these symposia shall continue for around ten years.

The more the number of participants increases, the more the substance of the symposium becomes vague and the perfume is gone. Therefore I beg your best understanding of our basic philosophies and thoughts.

Kyoto, the 1200-year-old capital of Japan, is widely known as a veritable treasure house of cultural and historical interest in Japan. It is, in truth, a city sitting in a nest of great natural beauty. I do hope you will enjoy your short stay in Kyoto.

In conclusion, I sincerely hope that all the participants, both Japanese and foreign, will return home bearing heartwarming memories of the symposium and a pleasant stay.

Thank you.

Participants in the Taniguchi International Symposium, November 3-8, 1981

Seated (left to right)

G. Toulouse P. Wölfle P. A. Lee F. Wegner D. J. Thouless R. Kubo Y. Imry Mrs. Imry

Second row (left to right)

Y. Ono H. Fukuyama K. Hoshino A. MacKinnon A. Kawabata S. Maekawa H. Kamimura Y. Nagaoka

Third row (left to right)

S. Hikami S. Yoshino T. Ando F. J. Ohkawa K. Niizeki D. Yoshioka F. Yonezawa H. Katagiri (secretary).

Contents

Part I. Anderson Localization

Part I

Anderson Localization

Survey of Localization: are Noninteracting Electrons Fully Understood?

D.J. Thouless

Department of Physics, FM-15, University of Washington
Seattle, WA 98195, USA

Abstract

The theory of localization of noninteracting electrons in a random potential is surveyed. The older perturbative approaches from the strong localization and weak scattering limit, the nonlinear sigma model, the self-consistent method, and scaling theory are summarized. The connection between spectral properties and localization is mentioned. The extent to which real and numerical experiment can confirm or conflict with the theory is discussed.

1. Introduction

It is now twenty-three years since ANDERSON [1] published his paper on the absence of diffusion in certain random lattices, and fifteen years since MOTT [2] and COHEN, FRITZSCHE and OVSHINSKY [3] stimulated a revival of interest in the subject of localization of electrons by a disordered potential. It is however only in the last two or three years that a coherent picture has begun to emerge, but now many different approaches to the problem give us the same results. The subject has quite suddenly matured.

The justification for the effort which went into trying to understand noninteracting electrons in a static disordered potential was the hope that we could establish a model that was as clear a guide to the behavior of electrons in disordered systems as the Bloch theory is to the behavior of electrons in a regular lattice. We know that the Bloch theory has to be modified in several vital respects before it is applied to real systems, both by the effects of disorder and by many-body effects such as electron-phonon and electron-electron interactions. In the same way a good theory of electrons in disordered systems should be clear enough and unambiguous enough to serve as a starting point for the discussion of the vital modifications of the model due to band structure and many-body effects. I believe we now have such a theory, and I am excited at the prospect of hearing so many of the people who have arrived at the same solution by different methods describe their own approaches. I will try to give a broad picture of what I expect them to say. I can, I hope, do this dispassionately, since I have contributed rather little to this synthesis, although I have understood many of the components that went into it for a long time. We do not, of course, understand everything about noninteracting electrons, but we know how to answer a limited number of questions.

Although we have compared the theory of electrons in disordered systems with the Bloch theory, in fact it is much more closely analogous to and owes much more to the theory of critical phenomena. It is hardly surprising that today's solution did not emerge ten years ago, since the theory of critical phenomena was only just beginning to achieve its own synthesis then. Science proceeds by the successive correction of errors, and so I am not distressed by how little of the work that was done in the late sixties and early seventies looks correct in the light of recent work. In particular I regard the concept of minimum metallic conductivity as one of the creative errors that helped the progress of science. It is interesting to reflect on the fact that there is more experimental evidence for minimum metallic conductivity [4] than for almost any of the other novel ideas that have emerged (except for variable range hopping), yet hardly any theorist would accept it.

2. Perturbative Methods

Anderson's original approach to the problem [1] was based on the application of perturbation theory to a system which was strongly localized by strong disorder. In some ways it resembles the URSELL-MAYER [5,6] approach to phase transitions by examining the convergence of a high temperature perturbation series. The arguments presented by Anderson showed convincingly that states are localized by strong disorder and that such states are quite distinct from the extended states familiar from the theory of metals, so that something like a phase transition must occur. The theory, however, predicted a much narrower range of stability for localized states than numerical simulations showed. The discrepancy was much larger than usually exists between mean field and more precise calculations of critical points, and it is still not really resolved. Critical exponents were calculated by using an analogy with the nonintersecting random walk problem [7,8]; although at least one of these exponents was (temporarily) experimentally confirmed [9], few people would now regard such calculations as close to the truth. Numerous attempts to improve on and extend the original work led to very little which has survived. One concept that did emerge was that a sort of block scaling argument could be used to map a general class of problems onto a version of the Anderson model.

The opposite approach, from the limit of weak disorder, proved to be more difficult in many ways. EDWARDS and a number of other people [10-12] developed functional integral techniques based on treating the disorder as a perturbation of free electrons. Some of the techniques that are used in recent field theory approaches were anticipated in this work, but few firm results came out of it, except for some predictions of critical exponents that agreed, more or less, with those obtained from the strong disorder limit, but which are also probably wrong. MOTT [4,13] also used quite different arguments about how disorder affects free electrons. He used the Ioffe-Regel criterion, which states that the mean free path, the distance over which the electron wave loses phase coherence, cannot be significantly less than a wavelength, to argue that metallic conductivity cannot be less than a certain value. When the conductivity reaches its minimum metallic value the electrons become localized and the conductivity drops sharply to zero. One of the most exciting predictions of this work, confirmed by experiment, was the existence of a universal value of the minimum metallic conductance per square in two dimensions [4,14]. Unfortunately this beautiful result survived for only three years before it was superseded by new theories and new experiments.

Recently various groups have calculated conductivity in the weak scattering limit of perturbation theory [15-17]. This has involved among other things a reinterpretation of the old results of LANGER and NEAL [18], and it is one of the vital ingredients of the scaling theory of localization. Among other things it predicts that there is no metallic conduction in two dimensions in the absence of magnetic effects.

3. Field Theory and Self-Consistent Theories

The success of methods based on a field theory formulation of critical phenomena suggested that such a formulation might be equally useful for the study of the effects of disorder on electron systems. It has been known for some time that the replica trick leads to a formulation in which the Lagrangian looks like the Lagrangian for a Landau-Ginzburg system with a spin variable with n degrees of freedom, in which n tends to the limit zero [19]. In this respect it is like the dilute polymer or nonintersecting random walk problems, but there is a vital difference in the sign of the quartic term. The usual symmetry breaking of a Landau-Ginzburg theory, in which one component of spin has a nonzero expectation value in the low temperature phase, has disastrous consequences, and it is therefore necessary to transform from the vector variable S_i^{α} to a tensor variable $Q_{ij}^{\alpha\beta} = S_i^{\alpha}S_j^{\beta}$ before further progress can be made [20]. A complication of this approach is that the existence of a nonzero density of states implies a nonzero value of the imaginary part of the Green function, and this implies a broken symmetry. The symmetry-breaking field is the imaginary part of the energy variable, and as this goes to zero the symmetry remains broken within the energy bands, which in a model with Gaussian disorder includes the whole of the real axis. This broken symmetry is associated with nonzero expectation values for the expectation value of diagonal elements of Q. The transition between localized and extended states is more subtle. Since it is related to the conductivity it can be studied by looking at the Q-Q correlation functions. The metallic phase is one in which the correlation falls off with a simple power law of distance. This is what should be expected if there is a Goldstone mode associated with the broken symmetry [21]. The mobility edge is signalled by the disappearance of this Goldstone mode. This may be somewhat analogous to what happens in the commensurate-incommensurate phase transitions that have attracted a lot of attention recently [22]. A phase which is floating on a substrate has a spectrum which goes down to zero frequency, while a phase locked to the substrate has no Goldstone modes and consequently can have defects. The lower critical dimension for this is 2, and the region of $2+\varepsilon$ dimensions has been explored in detail [20,23]. The scaling properties found are strikingly similar to those found by other methods. I hope to learn more about these matters from the talks of Wegner and Hikami.

A method which seems simpler to someone brought up on the traditional formulations of many-body theory involves a self-consistent resummation of Feynman diagrams, and has been developed by VOLLHARDT and WÖLFLE [24] and by KAWABATA [25]. As in the weak scattering work mentioned earlier the propagation of an electron is assumed to be diffusive over distances somewhat greater than the mean free path. A self-consistent interpolation is possible between the weak scattering and the strongly localized regimes, and this agrees remarkably well with the results obtained from the critical properties of the nonlinear sigma model. This we shall also hear more about.

4. Scaling and Conductance

Although a number of us had been struggling to apply the concept of scaling to localization problems for some years, it was not until the work of ABRAHAMS, ANDERSON, LICCIARDELLO and RAMAKRISHNAN [15] that we had a theory clear and direct enough to be useful. This was based on the idea that consideration of blocks of material leads to a rescaled Anderson model in which the coupling between sites is replaced by a coupling between blocks, and the only parameter of importance is the ratio of the coupling between these blocks to the spacing of energy levels within the blocks [26]. Furthermore, this ratio determines the conductance of a block (actually it is the square of this ratio rather than, as I originally stated, the ratio itself that is proportional to conductance[27]). Since localization is understood in the strong disorder limit and the conductance can be calculated in the weak disorder limit a reasonable interpolation between the two is almost immediately available.

In one dimension scaling theory gives the well-known result, first obtained by MOTT and TWOSE [28] and by LANDAUER [29], that all states are localized. In two dimensions it says that all states are also localized, which was contrary to what most of us thought before, although there were already indications that this was so. In three dimensions it said that there is no minimum metallic conductivity, since as the conductance on small length scales approaches the critical value the resistivity on large length scales increases without limit.

The finite size scaling calculations of PICHARD and SARMA [30] and of MACKINNON and KRAMER [31] add further depth to our understanding of scaling. Since I disagree with the interpretation of one pair of authors and like that of the other pair it is fortunate MacKinnon is here to explain what they have done. The idea of a finite size scaling theory is that a strip or wire of finite cross section is made long enough that one-dimensional localization dominates. The ratio of the one-dimensional localization length ℓ_M to the diameter M is a function, at least in the critical region, of the ratio of some characteristic length ℓ_∞ to M. The length ℓ_∞ is the bulk (or two-dimensional) localization length for the wire (or strip) when there is bulk (or two-dimensional) localization, and it is related to the conductivity σ by $\ell_\infty = e^2/h\sigma$ in the extended regime. It is therefore given by the length scale at which coupling between blocks is comparable with energy spacing. In two dimensions ℓ_∞ seems to go exponentially to infinity as the disorder goes to zero, but in three dimensions it diverges at a finite value of ℓ_M/M. If the relation between ℓ_M/M and conductance is known in one dimension, then the behavior of ℓ_M/M as a function of disorder can be related to the behavior of the conductance as a function of disorder. This work allows scaling theory to be extended to shapes which are not simply cubes or squares, and allows anisotropy to be included.

5. Spectral Properties and Localization

Although it is clear that the spectrum averaged over an infinite system or over an ensemble does not give adequate information about localization, there is an intimate connection between the local, nonaveraged, properties of the spectrum, as given for example by the spectral representation of a diagonal element of the one-particle Green function, and localization. If the spectrum is continuous in some energy range, states in that energy range are not normalized and therefore extended, while a point spectrum corresponds to localized states [32]. If, as is generally believed but not proved, metallic conductivity and diffusion are associated with an absolutely continuous spectrum, what transport properties are associated with the singular continuous spectrum, where the density of states is zero almost everywhere, but there are no δ-functions? The support of such a spectrum has to be a Cantor set, nondenumerable but with measure zero. AZBEL [33] raised this question and gave some answers some years ago. It is of some physical interest, since two-dimensional electrons in a uniform magnetic field in a periodic potential whose unit cell encloses an irrational number of flux quanta have a singular continuous spectrum [34]. Dr. Kohmoto and I have recently started looking at this problem, but we have not yet come to a conclusion. There are some interesting scaling properties that suggest the AC conductivity $\sigma(\omega)$ may go to zero linearly with ω instead of going quadratically to zero, as with localized states, or to a constant, as with extended states.

6. Experimental and Numerical Results

Since the theory I have been describing is an attempt to describe accurately the properties of a well-defined but oversimplified model, comparison with experiment can be suggestive and revealing but never conclusive. We know that in order to use scaling theory we have to take into account the effects that limit the applicability of the theory. This problem was brought into focus most clearly by the prediction that any wire with a zero temperature resistance greater than a few tens of kilohms would have a resistance increasing linearly with length [35]. The zero-temperature scaling theory only applies for length scales less than those set by inelastic scattering of electrons, by magnetic fields, or by alternating electric fields. Other many-body effects may also be important. Some recent experiments, particularly on MOSFETs, give good agreement with scaling theory [36]. I like to argue that the puzzling results with thin wires [37] tell us we have not understood inelastic scattering mechanisms properly rather than that scaling theory is wrong, while the experiments that show minimum metallic conductivity really show nothing of the sort, but result from the high temperature of the electrons determining a rather short scaling length [4,38]. It is however time to think about how to measure three-dimensional exponents for the localization problem.

On the other hand some experiments are so clean that they present an immediate challenge to theorists. The discovery of quantized Hall effect in MOSFETs to an accuracy of at least 1 in 10^5 by KLITZING, DORDA AND PEPPER [39] had to be understood. In fact a number of arguments have already shown that this should have been expected [40]. It is clear that in a strong magnetic field there must be some extended states in each Landau level.

Numerical calculations also have to be interpreted with some caution. Most of us have used our calculations at one time or another to support an idea which we now think is incorrect. Numerical results properly obtained and analyzed can be used to falsify a theory or establish the numerical parameters of a correct theory.

7. Conclusions

I believe we have now reached a good level of understanding of localized electrons and of the transition from the localized to the extended regime. There is still detailed work to be done, but I do not doubt that most of the obvious problems will have a reasonably satisfactory solution in a year or so.

The problems of applying these ideas to complicated situations such as occur in granular metals or to doped semiconductors or superconductors, are different in scale, and that is where the real work is to be done.

There is also the question of how the insights obtained in this work can be applied to other problems. I would have been particularly interested to hear what Anderson wanted to say about the connections between localization and the spin glass problem. Various people have pointed out a connection between localization and the two-dimensional planar spin model, but I am not convinced by any of the arguments I have seen. I have already mentioned the connection between the localization problem and commensurate-incommensurate phase transitions which has been explored by AUBRY and ANDRE [22].

This work was supported in part by the National Science Foundation, grant no. DMR79-20785A01.

References

1 P.W. Anderson: Phys. Rev. 109, 1492 (1958)
2 N.F. Mott: Adv. Phys. 16, 49 (1967)
3 M.H. Cohen, H. Fritzsche, S.R. Ovshinsky: Phys. Rev. Lett. 22, 1065 (1969)
4 N.F. Mott, M. Pepper, S. Pollitt, R.H. Wallis, C.J. Adkins: Proc. R. Soc. A 345, 169 (1975)
5 H.D. Ursell: Proc. Cambridge Phil. Soc. 23, 685 (1927)
6 J.E. Mayer: J. Chem. Phys. 5, 67 (1937)
7 P.W. Anderson: Proc. Nat. Acad. Sci. 69, 1097 (1972)
8 T. Lukes: J. Noncryst. Solids 8-10, 470 (1972)
9 M. Pepper, S. Pollitt, C.J. Adkins: J. Phys. C 7, L273 (1974)
10 S.F. Edwards: J. Phys. C 3, L30 (1970)
11 R.A. Abram, S.F. Edwards: J. Phys. C 5, 1183 (1972); J. Phys. C 5, 1196 (1972)
12 K.F. Freed: Phys. Rev. B 5, 4802 (1972)
13 N.F. Mott: Phil. Mag. 22, 7 (1970)
14 D.C. Licciardello, D.J. Thouless: Phys. Rev. Lett. 35, 1435 (1975)
15 E. Abrahams, P.W. Anderson, D.C. Licciardello, T.V. Ramakrishnan: Phys. Rev. Lett. 42, 673 (1979)
16 B.L. Altshuler, A.G. Aronov: Solid State Commun. 36, 115 (1979)
17 B.L. Altshuler, P.A. Lee, D. Khmelnitskii, A.I. Larkin: Phys. Rev. B 22, 5142 (1981)
18 J.S. Langer, T. Neal: Phys. Rev. Lett. 16, 984 (1966)
19 S.F. Edwards: J. Phys. C 8, 1660 (1975)
20 F. Wegner: Phys. Reports 67, 15 (1980); see also paper at this Symposium
21 A.J. McKane, M. Stone: Ann. Phys. (NY) 131, 36 (1981)
22 S. Aubry, G. André: in Nonlinear Structure and Dynamics in Condensed Matter (Springer-Verlag, Berlin, 1978), p. 133
23 S. Hikami: Phys. Rev. B 24, 2671 (1981); see also paper at this Symposium
24 D. Vollhardt, P. Wölfle: Phys. Rev. B 22, 4666 (1980); see also paper at this Symposium
25 A. Kawabata, Solid State Commun. 27, 1393 (1981)
26 D.J. Thouless: in Ill-Condensed Matter, ed. R. Balian, R. Maynard and G. Toulouse (North-Holland Publishing Co., Amsterdam, 1979), p. 1
27 D.W. Anderson: Phys. Rev. B 23, 4828 (1981)
28 N.F. Mott, W.D. Twose: Adv. Phys. 10, 107 (1960)
29 R. Landauer: Phil. Mag. 21, 863 (1970)
30 J.L. Pichard, G. Sarma: J. Phys. C 14, L127 (1981)
31 A. MacKinnon, B. Kramer, preprint; see also paper by A. MacKinnon at this Symposium
32 H. Kunz, B. Souillard: Commun. Math. Phys. 78, 201 (1980)
33 M. Ya. Azbel: Phys. Rev. Lett. 43, 2954 (1979)
34 D.R. Hofstadter: Phys. Rev. B 14, 2239 (1976)
35 D.J. Thouless: Phys. Rev. Lett. 39, 1167 (1977)
36 J. Dynes: Physica B+C, to be published
37 J.T. Masden, N. Giordano: Physica B 107, 1 (1981)
38 N.F. Mott: Phil. Mag. 44, 265 (1981)
39 K. von Klitzing, G. Dorda, M. Pepper: Phys. Rev. Lett. 45, 494 (1980)
40 H. Aoki, T. Ando: Solid State Commun. 38, 1079 (1981); see also various contributions to this Symposium

The Anderson Transition and the Nonlinear σ-Model

F.J. Wegner
Institut für Theoretische Physik, Ruprecht-Karls-Universität
D-6900 Heidelberg, Fed. Rep. of Germany

1. Introduction

A particle (e.g. an electron) moving in a random one-particle tight-binding potential

$$H = \sum_{r,r'} v_{r,r'} |r><r'| \tag{1}$$

may have localized and extended eigenstates depending on the energy of the particle. The energy E_c which separates the localized states from the extended states is called the mobility edge. Extended states can carry a direct current whereas localized states are bound to a certain region and can move only with the assistance of other mechanisms (e.g. by phonon-assisted hopping). Thus the residual d.c. conductivity is expected to vanish for Fermi energies E in the region of localized states, and to be nonzero for E in the region of the extended states.

The ket $|r>$ in (1) stands for an atomic orbital at site r. We assume these orbitals to be orthonormal. The matrix elements $v_{r,r'}$ are random variables (with independent distributions or short-range correlations). In amorphous materials r itself will be the random position of an ion. Here we consider only a periodic Bravais lattice of sites r. Here the emphasis will be on the underlying symmetries of the problem, the mapping on a field theoretic model of interacting matrices and the consequences for the behaviour in two and 2 + ε dimensions. Besides the lectures on disordered systems given in this volume and the original literature I also refer to lectures given in Brasov 1979 [1] and Les Houches 1980 [2].

2. Mapping on a Static Problem and Continuous Symmetry

Introducing energies $z_1 = E_1 + \omega/2$ and $z_2 = E - \omega/2$ with real E and Im $\omega > 0$ one defines the advanced and retarded Green's functions

$$\mathcal{G}(r,r', E \pm \omega/2) = <r| \frac{1}{E \pm \omega/2 - H} r'> \tag{2}$$

for the Hamiltonian (1). It can be rewritten [3,4] (for symmetric v)

$$\mathcal{G} = \mp i <S^{\pm}(r)\, S^{\pm}(r')> \tag{3}$$

where the average is defined by

$$<A(S)> = \prod_{r,\pm} \int_{-\infty}^{+\infty} dS(r) A(S)\, e^{-H(S)}/Z \tag{4}$$

with the curly Hamiltonian

$$\mathcal{H} = \sum_{rr'} \frac{\mp i}{2} S^{\pm}(r)((E \pm \omega/2)\delta_{rr'} - v_{rr'}) S^{\pm}(r'). \tag{5}$$

The contribution $-i\omega S^2/4$ guarantees the convergence of the integrals in (4).

The partition function Z depends on v. One gets rid of the denominator Z by introducing m replicas $S_a, a = 1, \ldots, m$. Then the denominator reads Z^m, which with the choice m = 0 yields unity. Thus one finds the averaged one-particle

$$\begin{aligned} G(r,r', E \pm \omega/2) &= \overline{\mathcal{G}(r,r', E \pm \omega/2)} \\ &= \mp i \langle S_a^{\pm}(r)\, S_a^{\pm}(r')\rangle \end{aligned} \tag{6}$$

and two particle Green's function

$$\begin{aligned} K(r_1, r'_1, r_2, r'_2) &= \overline{\mathcal{G}(r_1, r'_1, E + \omega/2)\mathcal{G}(r_2, r'_2, E - \omega/2)} \\ &= \langle S_a^{+}(r_1)\, S_a^{+}(r_1')\, S_b^{-}(r_2)\, S_b^{-}(r_2')\rangle \end{aligned} \tag{7}$$

with

$$\langle A(S)\rangle = \prod_{r,\pm,a} \int_{-\infty}^{+\infty} dS_a^{\pm}(r)\; A(S_b)\; \overline{e^{-\mathcal{H}(S)}} \tag{8}$$

and

$$\mathcal{H} = \mathcal{H}_0 + \mathcal{H}' \tag{9}$$

$$\mathcal{H}_0 = -\frac{i}{2} \sum_{rr'} (E\delta_{rr'} - v_{rr'}) \sum_a (S_a^{+}(r)\, S_a^{+}(r') - S_a^{-}(r)\, S_a^{-}(r')) \tag{9a}$$

$$\mathcal{H}' = -\frac{i\omega}{4} \sum_{ra} \left[(S_a^{+}(r))^2 + (S_a^{-}(r))^2\right]. \tag{9b}$$

Apparently, $\mathcal{H}_0$ is invariant under global pseudoorthogonal transformations which leave

$$\sum_a (S_a^{+}\, S_a^{+\prime} - S_a^{-}\, S_a^{-\prime}) \tag{10}$$

invariant. Thus in the zero frequency limit $\omega \to 0$ $\mathcal{H}$ obeys an O(m,m) symmetry with m = 0. This continuous symmetry is broken by $\mathcal{H}'$. Here $-i\omega$ plays the role of the symmetry breaking field (in analogy to a magnetic field for an isotropic ferromagnet) and

$$\langle S_a^{+}(r)\, S_a^{+}(r) + S_a^{-}(r)\, S_a^{-}(r)\rangle = i(G(E + \omega/2) - G(E - \omega/2)), \tag{11}$$

which is proportional to the density of states for $\omega \to 0$, plays the role of the order parameter (in analogy to the magnetization).

3. Local Gauge Invariance and Composite Variables

Although there is not complete agreement many authors [4-7] assume that the averaged one-particle Green's function decays rapidly with increasing $|r - r'|$ both for localized and extended eigenstates. Due to the fluctuations in the potential the states in different systems lose phase coherence. Thus on a scale large in comparison to the phase coherence length the contributions of the eigenfunctions ψ_i of different systems cancel which yields the decay of

$$G(r,r',z) = \sum_i \psi_i^+(r)(z - E_i)^{-1} \psi_i(r'). \tag{12}$$

Here $\psi_i(r)$ and E_i are the eigenfunctions and energies of the Hamiltonian H. Similarly one expects the averaged two-particle Green's function K to become negligible unless the sites are pairwise close to each other.

Therefore on a length scale large in comparison to this phase coherence length the averaged Green's functions are invariant under arbitrary changes of the phase of $\psi_i(r)$ (gauge transformations). One expects the ensemble to be asymptotic (large distance) gauge invariant.

Obviously this suggests considering models which are a priori gauge invariant [7]. Such models can easily be constructed. If all matrix elements of v are independently distributed with the restriction that v is Hermitian then it suffices to require a symmetric distribution. If we change the sign of the eigenfunction at orbital r then this new state is the eigenfunction of the Hamiltonian whose matrix elements $v_{rr'}$ and $v_{r'r}$ change sign for fixed r (with the exception of v_{rr}). Since this new Hamiltonian has the same probability as the original one, G, (12) vanishes unless r = r'.

Since the averaged Green's functions vanish in the local gauge model unless the points r coincide pairwise, all non-vanishing correlations can be expressed as correlations of the local composite variables

$$Q_{ab}^{pp'}(r) = i(pp')^{1/2} S_a^p(r) S_b^{p'}(r) \tag{13}$$

(with p,p' = ±1). Such a variable Q has already been introduced by Aharony and Imry [6] for p' = -p. In actual calculations one replaces the substitution (13) by a transformation from SS to Q [8,9]. For Gaussian distributed v's the correlations of factors SS and of factors Q are basically the same. Then the density of states ρ and the two-particle Green's functions can be expressed as

$$\rho(r) \sim \lim_{\omega \to 0} \langle Q_a^{++}(r) - Q_{aa}^{--}(r)\rangle \tag{13}$$

$$K(q) = \sum_{r'} K(r, r', r', r)\, e^{iq(r-r')} \sim \langle Q_{ab}^{+-}(q)\, Q_{ba}^{-+}(-q)\rangle \tag{14a}$$

$$K''(q) = \sum_{r'} K(r, r, r', r')\, e^{iq(r-r')} \sim \langle Q_{aa}^{++}(q)\, Q_{bb}^{--}(-q)\rangle\ . \tag{14b}$$

The behaviour near the moblity edge is described by the critical behaviour of the variables Q.

4. Density of States, Scaling

The symmetry breaking term (9b) yields an orientation of the "quadrupole" Q,

$$\langle Q_{ab}^{pp'}(r)\rangle = \delta_{pp'}\, \delta_{ab} G(r, r, z_p). \tag{15}$$

We expect [10] that Q has m(=0) eigenvalues $\lambda_1 \sim G(r, r, z_1)$ and m(=0) eigenvalues $\lambda_2 \sim G(r, r, z_2)$. We assume that the fluctuations of the eigenvalues of Q are irrelevant similar to the fluctuations of the length of the vector in the n-vector model. Thus we restrict Q to the set of matrices with eigenvalues λ_1 and λ_2 (zero-fold degenerate) which are obtained from the diagonal representation by means of pseudoorthogonal transformations. Thus Q obeys

$$(Q - \lambda_1)(Q - \lambda_2) = 0. \tag{16}$$

For small elongations Q^{12}, Q^{21} and $Q^{pp} \approx \lambda_p$ one obtains from (16)

$$Q^{11} = \tfrac{1}{2}(\lambda_1 + \lambda_2) + \left[(\tfrac{1}{2}(\lambda_1 - \lambda_2))^2 - Q^{12}Q^{21}\right]^{1/2} \tag{17}$$

and similarly for Q^{22}. Thus the "longitudinal" components Q^{11}_{ab}, Q^{22}_{ab} can be expressed in terms of the "transversal" components Q^{12}_{ab}, Q^{21}_{ab}. The matrix elements Q^{12}_{ab} are independent. The Taylor expansion of (17)

$$Q^{11}_{ab} = \lambda_1\delta_{ab} - \frac{1}{\lambda_1 - \lambda_2} \sum_c Q^{12}_{ac}Q^{21}_{cb} + \ldots \tag{18}$$

yields

$$\langle Q^{11}_{ab}\rangle = \lambda_1\delta_{ab} \tag{19}$$

since the summations run over m = 0 equal contributions and thus do not contribute. This tells us that

$$G(r, r, z_p) = \lambda_p \tag{20}$$

holds. The transverse fluctuations do not affect the averaged Green's function which thus stays constant at the mobility edge (critical exponent β = 0). This is particular for the m = 0 component problem. We now use the well-known results from conventional critical phenomena. We obtain the analogies to, e.g., an isotropic ferromagnet at temperature T and magnetic field h with magnetization m, and transverse and longitudinal susceptibilities $\chi_\perp$ and $\chi_\parallel$, as listed below in the first two rows.

$E - E_c$	$\hat{=}$	$T - T_c$	1
$-i\omega$	$\hat{=}$	h	$\beta + \gamma = d\nu$
q	$\hat{=}$	q	ν
ρ	$\hat{=}$	m	$\beta = 0$
$K(q)$	$\hat{=}$	$\chi_\perp(q)$	$-\gamma = -d\nu$
$K''(q)$	$\hat{=}$	$\chi_\parallel(q)$	$-\gamma = -d\nu$

The third column shows how these quantities scale, that is if $E - E_c$, $-i\omega$, and q are multiplied by factors b^1, $b^{\beta+\gamma}$, b^ν, then ρ , K and K" are rescaled by factors b^β, $b^{-\gamma}$, and $b^{-\gamma}$, resp. Since β = 0, and the scaling law

$2\beta + \gamma = d\nu$ holds, all exponents can be expressed by ν. Thus, e.g.

$$K(b^{\nu}q,\ b^{d\nu}\omega,\ b(E - E_0)) = b^{-d\nu}K(q,\omega\ ,\ E - E_c)\ . \tag{21}$$

The conductivity σ can be expressed by the diffusion constant D and K as

$$\sigma \sim \rho D \sim \rho^2\ \partial K^{-1}/\partial q^2; \tag{22}$$

thus it scales like $\sigma \sim b^s$ with

$$s = 2\beta + \gamma - 2\nu = (d-2)\nu. \tag{23}$$

All these scaling laws agree with the predictions from real-space renormalization for the homogeneous fixed point ensemble [5].

5. Renormalizable Field Theory, Universality Classes

Let us construct a Lagrangian L_{eff} for the matrices Q. Any local potential of Q obeying the full pseudo O(m,m) symmetry is a function of λ_1 and λ_2 only and thus a constant. Any interaction containing one gradient violates inversion symmetry and can be expressed as a surface integral. The most simple interaction containing two gradients is

$$L^0_{eff} = \tfrac{1}{4}K \int d^dr\ tr(\nabla Q(r)\nabla Q(r)). \tag{24}$$

It is the only interaction containing two derivatives ∇ obeying the full symmetry. All other interactions with two ∇'s can be reduced to (24) by means of (16). Contributions with more than two ∇'s turn out to be irrelevant near d = 2. The symmetry breaking term has to be added

$$L_{eff} = L^0_{eff} + \frac{\omega}{4v} \sum_a \int d^dr(Q^{11}_{aa}(r) - Q^{22}_{aa}(r)). \tag{25}$$

This model is very similar to the nonlinear σ-model on matrices with orthogonal symmetry. The "pseudo" symmetries yield the same diagrams; only some over-all-signs are flipped. Thus the ultra-violet divergencies can be removed by the same renormalizations. For given λ_1 and λ_2 the matrices Q are invariant under transformations of the subgroup O(m) x O(m). Thus the system is governed by the symmetry O(m,m)/O(m) x O(m). The W-function and thus the critical exponent $s = (d - 2)\nu$ has been calculated in $d = 2 + \varepsilon$ dimensions to two-loop [11] and three-loop order [12] for the model (25) with orthogonal symmetry yielding

$$s = 1 + O(\varepsilon^3). \tag{26}$$

This holds for spin-independent time-reversal invariant potentials. For a spin-dependent time-reversal invariant potential where isotropy is maintained in the average the unitary symplectic group Sp(0,0)/Sp(0) x Sp(0) applies [13,14]. If time-reversal invariance is broken but isotropy is maintained in the average then the unitary group U(0,0)/U(0) x U(0) governs the system yielding [14,15,12,8]

$$s = 1/2 + O(\varepsilon). \tag{27}$$

For two-dimensional systems the d.c. conductivity vanishes at all energies in the orthogonal and the unitary case [14-18]. In the symplectic case, however, there may be a region where the conductivity behaves better than ohmic

[13]. In all cases one finds σ of a square of length L in the quasi-metallic region

$$\sigma = \sigma_0 \frac{e^2\alpha}{\pi^2 h} \ln(\frac{L}{a}) + \dots \tag{28}$$

with α = 1,0, -1/2 in the orthogonal, unitary, and symplectic cases, respectively.

6. Density Fluctuations and Correlations

Density fluctuations can be investigated by introducing the average of the squared wave function within energy width

$$\rho_\eta(r,E) = \sum_i \frac{\eta/\pi}{(E-E_i)^2 + \eta^2} |\psi_i(r)|^2 \sim Q^{++}_{aa}(r) - Q^{--}_{aa}(r). \tag{29}$$

In one-loop order one finds for extended states (for the orthogonal case) [19]

$$\overline{\rho^k_\eta} \sim (E - E_c)^{-\mu_k} \tag{30}$$

with

$$\mu_k = k(k - 1) + O(\varepsilon). \tag{31}$$

This divergence indicates that the eigenfunctions show strong fluctuations in the amplitude near the mobility edge. In the region of localized states one obtains the average 2k-th power of the amplitude

$$\rho P_k = \lim_{\eta \to 0} \eta^{k-1} \rho^k_\eta \sim \overline{\sum_i |\psi_i(r)|^{2k} \delta(E - E_i)} \sim (E_c - E)^{\pi_k} \tag{32}$$

with

$$\pi_k = (k - 1)d\nu - \mu_k = (k - 1)(2\varepsilon^{-1} + 1 - k) + O(\varepsilon). \tag{33}$$

For d = 2 this reduces in the quasi metallic region to

$$P_k \sim \xi^{2-2k} |\ln\xi|^{k(k-1)} . \tag{34}$$

Using the idea of the short-distance expansion one may also estimate [20] the density correlations. In one-loop order the orthogonal case yields for extended states the r-dependence

$$\overline{\rho^k_\eta(0)\rho^{k'}_\eta(r)} \sim (\xi^{-\varepsilon} + r^{-\varepsilon})^{2kk'}.$$

Apparently ξ is the distance over which the amplitudes of the wavepackets are correlated. For localized states and $r \ll \xi$ one finds

$$\overline{\sum_i |\psi_i(0)|^{2k} |\psi_i(r)|^{2k'} \delta(E - E_i)} \sim (r^{-\varepsilon} - \xi^{-\varepsilon})^{2kk'} .$$

In two dimensions the r-dependence goes like $(\ln(\xi/r))^{2kk'}$. For sufficiently large ξ/r the decay is very slow and can be well approximated by a power law with an effective exponent for some finite region of r, which depends on ξ/r.

References

1 F. J. Wegner in International Summer School "Recent Advances in Statistical Mechanics", Brasov (1979) 63

2 D. J. Thouless, p.5, E. Abrahams, p.9, F. Wegner, p.15, Phys. Reports 67 (1980)

3 S. K. Ma, unpublished (1972); S. F. Edwards, J. Physics C8 (1975) 1660; A. Nitzan, K. H. Freed, M. H. Cohen, Phys. Rev. B15 (1977) 4476

4 D. J. Thouless, J. Phys. C8 (1975) 1803

5 F. J. Wegner, Z. Phys. B25 (1976) 327

6 A. Aharony, Y. Imry, J. Physics C10 (1977) L487

7 F. J. Wegner, Phys. Rev. B19 (1979) 783

8 L. Schäfer, F. Wegner, Z. Phys. B38 (1980) 113

9 A. B. Harris, T. C. Lubensky, Sol. St. Comm. 34 (1980) 343

10 F. Wegner, Z. Phys. B35 (1979) 207

11 E. Brézin, S. Hikami, J. Zinn-Justin, Nucl. Phys. B165 (1980) 528

12 S. Hikami, Progr. Theor. Phys. 64 (1980) 1466

13 S. Hikami, A. I. Larkin, Y. Nagaoka, Prog. Theor. Phys. 63 (1980) 707

14 R. Oppermann, K. Jüngling, Phys. Lett. 76A (1980) 449

15 K. Jüngling, R. Oppermann, Z. Phys. B38 (1980) 93

16 E. Abrahams, P. W. Anderson, D. C. Licciardello, T. V. Ramakrishnan, Phys. Rev. Lett. 42 (1979) 673

17 R. Oppermann, F. Wegner, Z. Phys. B34 (1979) 327

18 L. P. Gorkov, A. I. Larkin, D. E. Khmelnitzkii, Pis. Zh. Eksp. Teor. Fiz. 30 (1979) 248; JETP Lett. 30 (1979) 228

19 F. Wegner, Z. Phys. B36 (1980) 209

20 F. Wegner, to be published

Anderson Localization in Nonlinear σ Model Representation

S. Hikami

Research Institute for Fundamental Physics
Kyoto University, Kyoto 606, Japan

ABSTRACT

It is shown that the Anderson localization problem of electron impurity scattering is described by a special type of the field theoretical model, which is called a nonlinear σ model. The diagrammatic $1/E_F\tau$ expansion has one-to-one correspondence with small coupling t-expansion in the nonlinear σ model. The renormalization problems in the Anderson localization are solved by the equivalence of this problem to the nonlinear σ model, which is a renormalizable theory in two and three dimensions in the sense of small t and small $\varepsilon=d-2$ double expansions. The β-function is calculated up to three-loop order and it shows the asymptotic free behavior in two dimensions when the system has time-reversal and space-inversion invariance. The diffusion constant D is obtained as a function of frequency ω. The results of the unitary and the symplectic case are also presented. The lattice nonlinear σ model in one dimension is considered for the comparison with the β-function of the one dimension electron scattering problem.

1. Introduction

In this article, we will discuss a renormalization group treatment of the Anderson localization problem [1,2] in two dimensions and three dimensions. The summation of the ladder diagrams in electron impurity scattering is well known to give the diffusion propagator of electron in a random potential [3,4]. Beyond the simple ladder diagram, one can discuss the interaction of the diffusion mode by the perturbation of $1/E_F\tau$. The first correction arises from a maximum crossed diagram. And this diagram gives the logarithmic infrared divergence in two dimensions [5]. This logarithmic term indicates the possible applications of the renormalization group method. ABRAHAMS et al. [6] have derived the renormalization group β-function of the conductance g and they have argued that there is no metallic phase in two dimensions in one-loop order calculation. The renormalization problem has been studied in [7,8]. Also starting from a field theoretical model, the same result was obtained by 1/N expansion [9].

It has been considered that only localized phases exist in two dimensions. Extended state may correspond to ordered phase, and localized state is a disordered phase. The problems of Anderson localization which we will discuss are how to describe the phase transition between extended state and localized state. WEGNER [10] has conjectured that this extended and localized states phase transition is described by a special type of a nonlinear σ model or, in other words, that Anderson localization problem belongs to the same universal class as the nonlinear σ model, and he has argued that the same critical exponents and the same renormalization procedure hold.

The renormalization problem of the Anderson localization will be solved if one establishes the equivalence of the localization problem to some field theoretical models like a renormalizable nonlinear σ model [11]. In two dimensions, in addition to an ultraviolet divergence, there exists an infrared divergence for a diffusion mode. We will show that the diffusion propagator corresponds to the transverse propagator of a nonlinear σ model and the external frequency ω plays the role of symmetry breaking field and the infrared regulator [12]. Although it is not necessary to go back to the field theoretical model and it is sufficient to consider the renormalization procedure of Anderson localization problem directly, the comparison with nonlinear σ model is very fruitful in many respects. Especially one can avoid some confusion of the β-function which exist in literature [6-8] by following the procedure of the nonlinear σ model. The nonlinear σ model treats the dynamics of Goldstone modes depending on internal symmetries. The universal quantities like critical exponents are related to the internal symmetries [13-15]. By the symmetrical argument, we can obtain the β-functions up to three-loop order [16-18] which can be applied to Anderson localization problem. This is practically important. In addition to the technical advantage, the representation by the nonlinear σ model gives the ω dependence of the diffusion constant explicitly. We also briefly discuss the nonperturbative analysis by one dimensional lattice nonlinear σ model [19].

2. Diagrammatic expansion for electron impurity scattering

We will review the diagrammatic expansion for the interacting diffusion mode [8,12]. The Hamiltonian of the electron-impurity scattering is given by

$$H = \frac{p^2}{2m} + \sum_i V(x - X_i) \quad . \tag{1}$$

The second term is a random potential term which arises from the impurity located randomly at X_i. We write this Hamiltonian as

$$H = \sum_p \varepsilon_p a_p^+ a_p + \sum_p V(q)\rho_q \sum_p a_{p+q}^+ a_p \quad . \tag{2}$$

The quantity ρ_q is given by

$$\rho_q = \sum_j \exp(-i\vec{q}\cdot\vec{X}_j) \quad . \tag{3}$$

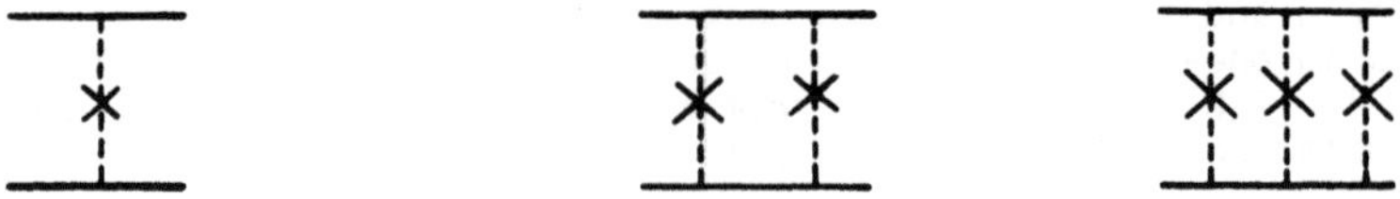

Fig. 1 Ladder diagram

We consider the two body Green function of density-density with Hamiltonian (2). The random average is easily done in diagrammatic expansion. The diagram should be connected before random field ρ_q is contracted. We assume that the ρ_q is a Gaussian random field. The ladder diagrams of two body Green function are shown in Fig.1. Two solid lines represent the retarded Green function $G^R(k,\omega)$ and advanced Green function $G^A(k,\omega)$ and a broken line represents ρ_q random field. The cross in a broken line represents the contraction by random average and the second Born scattering by an impurity is considered. The ladder summation gives the diffusion propagator as

$$D(q,\omega) = (1/\tau)[1 - \Pi(q,\omega)/\tau]^{-1} = \tau^{-2}[D_0 q^2 - i\omega]^{-1} \tag{4}$$

where $\Pi(q,\omega)$ is polarization part and it is calculated as

$$\Pi(q,\omega) = \int G^R(p+q,\ \omega+E_F)G^A(p,E_F)d^dp$$
$$= \tau - (v_F{}^2/d)q^2\tau^3 + i\omega\tau^2. \tag{5}$$

The retarded and advanced Green functions are given (+ : retarded, - : advanced)

$$G^{R,A}(k,\omega) = \frac{1}{\omega - \frac{1}{2}k^2 \pm \frac{i}{2\tau}} \quad . \tag{6}$$

The energy of electron is assumed to be Fermi energy E_F (v_F is Fermi velocity). The relaxation time is noted by τ and the mass is set equal to one. The diffusion constant D_0 of the electron is given by the coefficient of q^2 term in (4) as

$$D_0 = \tau\, v_F{}^2/d = E_F\tau \quad . \tag{7}$$

If D_0 is large enough, one can perform the systematic $1/E_F\tau$ expansion as usual loop expansions. The internal momentum integrations of the diagrams are easily done by the fact that the energy of the particle is fixed to Fermi energy and nonvanishing diagrams consist of proper combination of retarded and advanced Green function. As shown in [12], the diagrams of one-loop order for the correction to diffusion propagator are given in Fig.2.

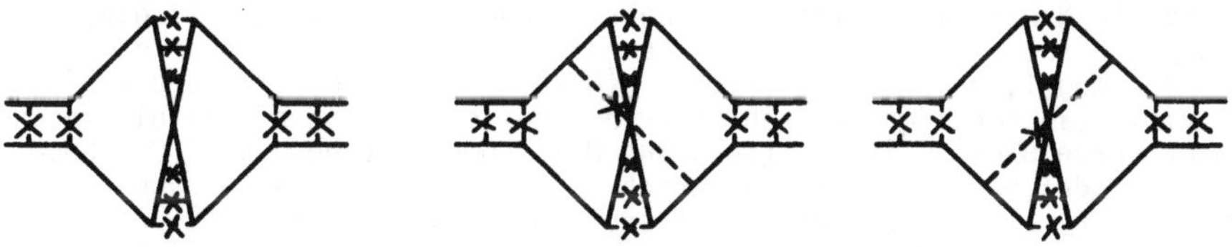

Fig. 2 One-loop correction

These diagrams contribute to the diffusion constant as

$$D = D_0(1 - \frac{\hbar}{2\pi E_F\tau}\ \ell n\ \frac{1}{\omega\tau}) \quad . \tag{8}$$

As shown in Fig.2 the one-loop diagram is obtained by the contraction of the four point vertex. Vertex part is obtained by cutting the legs of the ladder propagator. The two-loop diagrams have been investigated in [12] and there appears six point vertex. It is easy to calculate firstly the vertex part for small external momentum before doing loop integrals. The vertex parts turn out to be proportional to square of external momentum. Four point vertex part with external momentum $\vec{q}_1, \cdots \vec{q}_4$ (Fig.2) becomes

$$V_4(\vec{q}_1, \cdots, \vec{q}_4) = D_0 \tau^4 [(\vec{q}_1+\vec{q}_2)\cdot(\vec{q}_3+\vec{q}_4) + (\vec{q}_1+\vec{q}_4)\cdot(\vec{q}_2+\vec{q}_3)] \quad . \qquad (9)$$

The fact that the vertex has momentum square dependence indicates that the theory is highly nonlinear field theory which includes the current-current interaction. We will show that the theory becomes a special type of nonlinear σ model which has such momentum square, current-current interaction.

This diagrammatic loop expansion gives a systematic procedure for inverse of $E_F\tau$ expansion. We notice that the diagrams of the two particle Green function consist of one retarded Green function line and one advanced Green function line. Since the diagrams should be connected before the random average, one has only two different lines if one cuts whole crossed points in the diagrams. This situation is similar to a problem of polymer [20,21] or random self-avoiding walk. These random problems have been discussed by N=0 technique. One can avoid the random average procedure for logarithm of the partition function by this N=0 replica formulation. In Anderson localization problem, we can also apply this N=0 technique. The main difference to the polymer problem is that we need two different N=0 techniques due to independent retarded line and advanced line. In reference [10], one can also see the argument of replica method. The random average technique for Fermion is differently formulated by Grassmann variable in [15].

3. Nonlinear σ model

Before discussing the correspondence between the diagrammatic expansion of the Anderson model and the diagrammatic expansion of nonlinear σ model, we briefly review the general properties of the nonlinear σ model and its renormalization group procedure. The simplest nonlinear σ model is N-vector spin model where N represents the number of the components of spin. The spin takes the value on N-1 dimensional unit sphere S^{N-1}. This unit sphere S^{N-1} is a manifold represented by a coset space O(N)/O(N-1). The polymer case corresponds to N=0 limit of this nonlinear σ model. Demanding U(1) gauge invariance for the nonlinear σ model, we have a different nonlinear σ model called CP^{N-1} model, which is defined on U(N)/U(N-1)×U(1) coset space [22]. Generally, we can consider the nonlinear σ model defined on a manifold which is a coset space G/H: G is Lie group and H is compact subgroup of G. Such a nonlinear σ model has a Hamiltonian which satisfies the global G symmetry and local H symmetry. Usual symmetries of group G are orthogonal O(N), unitary U(N) and symplectic Sp(N) classical group. The subgroup H takes also such three different types of groups or their products. The meaning of a coset space is following: If we have a relation between elements of group G,

$$g' = h\, g\, h^{-1} \qquad g,\, g' \in G\ , \qquad h \in H\ , \qquad (10)$$

we have equivalent relation between g' and g. This equivalence leads to classification of G by H. Although H is arbitrary subgroup of G, it is necessary that H is maximum compact subgroup of G for the renormalizable theory with one coupling constant. (When H is such maximum compact subgroup of G, the coset space G/H is called as symmetric space.) The dimensions of O(N), U(N) and Sp(N) are N(N-1)/2, N^2 and N(2N+1), respectively. For N-vector model defined on O(N)/O(N-1) manifold, the dimension of manifold G/H becomes dim G - dim H = N-1. Therefore, N-vector model describes N-1 Goldstone modes in low temperature region. If phase transition occurs, the O(N) symmetry is recovered in high temperature region. In general, nonlinear σ models are considered for such symmetry breaking phenomena. We also note that three ensembles, orthogonal, unitary and symplectic case, which are described by the symmetric spaces of U(N)/O(N), U(N) and U(2N)/Sp(N), have been considered for random matrix problem [23,24] and for fine particle theory [25] where spin-orbit interaction is represented by symplectic ensemble due to space-inversion breaking symmetry and magnetic field case is represented by unitary ensemble due to time-reversal breaking symmetry. Among symmetric space G/H-nonlinear σ models, we have a class named a Grassmannian nonlinear σ model which is defined on $O(N)/O(p)\times O(N-p)$, $SU(N)/S[U(p)\times U(N-p)]$ and $Sp(N)/Sp(p)\times Sp(N-p)$ manifolds. If G is noncompact case, global symmetry groups O(N), SU(N) and Sp(N) are replaced by O(p,N-p), SU(p,N-p) and Sp(p,N-p) noncompact groups.

The relation between Anderson localization and p=N=0 Grassmannian nonlinear σ model has been first pointed out by WEGNER [10]. He has predicted that the exponent of the conductivity in three dimensions is same as the exponent ν of that nonlinear σ model, which has been calculated by BRÉZIN, HIKAMI and ZINN-JUSTIN [1]. This relation has been also studied in [26-28].

The G/H-nonlinear σ model is a renormalizable theory in two dimensions by naive power counting analysis and is also renormalizable in three dimensions where ultra-violet stable fixed points exist, in the sense that double small coupling and small $\varepsilon=d-2$ expansions hold. In $d>2$, there exists low temperature ordered phase where Goldstone mode appears. In two dimensions, by the asymptotic free behavior of β-function, the ordered phase shrinks to zero coupling constant and the system becomes disordered state for finite coupling constant. This situation is very similar to the Anderson localization discussed in the previous section. It is desirable to find a proper parametrization of the noncompact Grassmannian nonlinear σ model, which gives the identification of transverse propagator (Goldstone mode) to the diffusion propagator.

The Grassmannian nonlinear σ model is described by a projection matrix P,

$$P = g\,\eta\,g^* \tag{11}$$

where

$$\eta_{ij} = \begin{cases} \delta_{ij} & 1 \le i,j \le p \\ 0 & p < i,j \le N \end{cases} \tag{12}$$

and g is a group element. This projection matrix satisfies

$$P^2 = P \quad . \tag{13}$$

The dimension of Grassmannian $O(N)/O(p)\times O(N-p)$ is dim O(N) - dim O(p) - dim O(N-p) = p(N-p). Therefore, if we introduce field matrix variable Q_{ij} ($i=1,\cdots p$; $j=1,\cdots N-p$), the projection matrix becomes

$$P = \begin{pmatrix} \frac{1}{2} + (\frac{1}{4} + QQ^+)^{1/2} & iQ \\ iQ^+ & \frac{1}{2} - (\frac{1}{4} + Q^+Q)^{1/2} \end{pmatrix} \quad . \tag{14}$$

This form is convenient for a noncompact case. (For a compact case, we replace Q by iQ.) The simplest renormalizable action density becomes

$$A = \frac{1}{4t} \operatorname{tr} \partial_\mu P \, \partial_\mu P \quad . \tag{15}$$

If we expand this action by the power series of the matrix Q, we have an infinite series expansion. This treatment is generalization of usual O(N)/O(N-1) nonlinear σ model where σ field and (N-1) component π-field satisfy $\sigma^2+\pi^2=1$ constraint and σ field is written by π-field. The action of (15) becomes

$$A = \frac{1}{2t}\{\operatorname{tr} \partial_\mu Q \, \partial_\mu Q^+ - \frac{1}{2}\operatorname{tr} \partial_\mu (QQ^+)\partial_\mu (QQ^+) - \frac{1}{2}\operatorname{tr}(\partial_\mu Q^+ Q)(\partial_\mu Q^+ Q)$$
$$+ \operatorname{tr} \partial_\mu (QQ^+)^2 \partial_\mu (QQ^+) + \operatorname{tr} \partial_\mu (QQ^+)\partial_\mu (QQ^+)^2 + \cdots \} \quad . \tag{16}$$

If we make Fourier transformation, we have

$$A = \frac{1}{2t} \operatorname{tr}\{\sum q^2 Q(q)Q^+(-q) + \frac{1}{2}\sum[(\vec{q}_1+\vec{q}_2)\cdot(\vec{q}_3+\vec{q}_4) + (\vec{q}_1+\vec{q}_4)\cdot(\vec{q}_2+\vec{q}_3)]$$
$$\times Q(q_1)Q^+(q_2)Q(q_3)Q^+(q_4) + \cdots \} \tag{17}$$

where Q^+ is Hermite's conjugate $Q^+=({}^t\bar{Q})$. In orthogonal case Q becomes the real $p\times(N-p)$ matrix.

The perturbation expansion about small t for the propagator <QQ> coincides with the loop expansion for diffusion propagator by taking p=N-p=0 limit. The coupling constant t corresponds to $1/E_F\tau$. The Q-field represents the Goldstone mode of the nonlinear σ model and the action of (16) is interpreted as an interaction of diffusion mode.

The nonlinear σ model is a renormalizable theory and if we introduce two renormalization constants Z_1 of coupling and Z of wave function, we can make the Green function be finite. There exists additional problem of infrared divergence in two dimensions. We use the external field to regularize this infrared divergence. This process has been discussed in O(N)/O(N-1) nonlinear σ model [29]. The minimal subtraction scheme in dimensional regularization method is particularly convenient to consider the renormalization problem of the nonlinear σ model. This method has been worked out in [11, 12]. If we consider the inverse of correlation function <QQ> (the two point vertex part, $\Gamma^{(2)}(q)$, this quantity has ultraviolet divergence which is renormalized by Z_1 and Z. According to minimal subtraction in dimensional regularization method, the renormalized coupling constant t is related to the bare coupling constant t_0 as

$$t_0 = t\,\mu^{2-d}\,Z_1 \tag{18}$$

where the momentum μ is a quantity which regularizes the renormalization. The field Q is also renormalized as

$$Q_0 = \sqrt{Z}\,Q \tag{19}$$

and the following symmetry breaking term is added to the action

$$-\frac{\sqrt{Z}}{4t}\,h \operatorname{tr}\sqrt{1 + ZQQ^+} \tag{20}$$

where the bare field h_0 is renormalized as

$$h_0/t_0 = h\mu^{d-2}/t\sqrt{Z} \quad . \tag{21}$$

The two-point vertex function $\Gamma^{(2)}(q,h)$ is expanded for small t as (ε=d-2)

$$\Gamma^{(2)}(q,h) = \frac{Z}{Z_1 t} q^2 + \frac{\sqrt{Z}}{t} h + \frac{N+2}{\varepsilon} q^2 + \frac{N}{\varepsilon} h + O(t) \quad . \tag{22}$$

According to the minimal subtraction scheme, $\Gamma^{(2)}$ becomes finite for $\varepsilon\to 0$ by

$$Z = 1 - \frac{2N}{\varepsilon} t + O(t^2), \quad Z_1 = 1 - \frac{N-2}{\varepsilon} t + O(t^2) \quad . \tag{23}$$

The renormalization group $\beta(t)$-function is defined by

$$\beta(t) = \mu\frac{\partial t}{\partial\mu}\Big|_{t_0} \quad . \tag{24}$$

From the fact that the bare coupling constant t_0 is independent of μ, we have

$$0 = \mu \frac{\partial \ell n t_0}{\partial\mu} = \mu\frac{\partial t}{\partial\mu}\frac{\partial}{\partial t} \ell n Z_1 + (2-d) + \frac{\mu}{t}\frac{\partial t}{\partial\mu} \quad . \tag{25}$$

Thus we get

$$\beta(t) = \varepsilon t/(1 + t\frac{\partial}{\partial t} \ell n\ Z_1) \quad . \tag{26}$$

Expanding $\beta(t)$ by the small t parameter, we have

$$\beta(t) = \varepsilon t - a_1 t^2 - a_2 t^3 - a_3 t^4 - \cdots \quad . \tag{27}$$

The coefficient a_1 is 2 and a_2 is zero [11]. Recent calculation also determines a_3 to be zero [12,17] for $O(p,N-p)/O(p)\times O(N-p)(p=N=0)$ nonlinear σ model. The β-function in two dimension has negative coefficient in one loop order. This leads to asymptotic freedom. Other renormalization constant Z becomes one which is studied up to two loop order, and it seems that there is no divergence in wave function renormalization constant Z in the limit $\varepsilon\to 0$. This is related to the fact that there is only elastic scattering in Anderson localization problem,and consequently energy is conserved, and no $\omega\ell n\omega$ term in a diffusion propagator.

We have the renormalization group equation for $\Gamma^{(2)}(q,h)$ from the fact that the bare $\Gamma^{(0)}$ is independent of μ as [29]

$$\{\mu\frac{\partial}{\partial\mu} + \beta(t)\frac{\partial}{\partial t} - \zeta(t) + [\zeta(t) + \frac{\beta(t)}{t} - (d-2)]h\frac{\partial}{\partial h}\}\Gamma^{(2)}(p,t,h,\mu) = 0 \tag{28}$$

where $\zeta(t)$ is defiend by

$$\zeta(t) = \beta(t)\frac{\partial}{\partial t} \ell n Z \quad . \tag{29}$$

Equation (27) is derived from the relation between bare and renormalized two point functions:

$$\Gamma_B^{(2)} = Z\, \Gamma_R^{(2)} \,. \tag{30}$$

We approximate β and ζ functions to be

$$\beta(t) = \varepsilon t - 2t^2 \quad , \quad \zeta(t) = 0 \quad , \tag{31}$$

and try to solve the diffusion constant D as a function of ω. The diffusion constant D is defined as a coefficient of q^2 term in $\Gamma^{(2)}$ as $\Gamma^{(2)} \sim D\mu^{d-2}q^2$. The renormalization group equation becomes in two dimensions,

$$\left(\mu\frac{\partial}{\partial\mu} - 2t^2 \frac{\partial}{\partial t} - 2t \frac{\partial}{\partial \ell nh}\right)D(t,\mu,h) = 0 \tag{32}$$

where $t=1/E_F\tau$ and $h=-i\omega t$. The field h is scaled as $h/\mu^2=x$. Then, the renormalization group equation is written as

$$\left(-2\frac{\partial}{\partial \ell nx} - 2t^2 \frac{\partial}{\partial t} - 2t \frac{\partial}{\partial \ell nx}\right)D(t,x) = 0 \quad . \tag{33}$$

In the small $1/E_F\tau$, we have perturbation expansion which has been discussed;

$$D(t,x) = \frac{1}{t} + \ell nx + O(t) \quad . \tag{34}$$

The solution D of (33) which satisfies this initial condition is written as

$$D \exp\left(D - \frac{1}{t}\right) = \frac{h}{t\mu^2} = - \frac{i\omega}{\mu^2} \quad . \tag{35}$$

We see that D becomes zero for vanishing ω. For small ω,

$$\mathrm{Im} D \simeq -\omega \quad , \ \mathrm{Re}\, D \simeq \omega^2 \quad . \tag{36}$$

The conductivity $\sigma(\omega)$ is related to this diffusion constant and for small ω,

$$\sigma \cong e^2\rho D s \tag{37}$$

where ρ is density of state and s is spin degeneracy.

For $d=2+\varepsilon$ dimensions, we have

$$\left\{\varepsilon + (\varepsilon t-2t^2)\frac{\partial}{\partial t} - (2 + 2t)\frac{\partial}{\partial \ell nx}\right\}D = 0 \quad . \tag{38}$$

The solution is expressed by a scaling function f as

$$D = \Delta t \ f(h/(\Delta t)^{1+2/\varepsilon}) \tag{39}$$

where $\Delta t=|t-t_c|/t_c$. From the analogy of the equation of state in a nonlinear σ model [29], the equation of state might be written as

$$\frac{h}{\tilde{D}^{d/(d-2)}} = \left(\frac{t}{t_c}\right)^{-\frac{2}{d-2}} \left(1 + \frac{(t/t_c) - 1}{\tilde{D}^{1/\nu(d-2)}}\right)^{\frac{2}{d-2}} \tag{40}$$

where $\tilde{D}=Dt$ (normalized diffusion constant). This equation is an approximated result by d-2 expansion. In the limit $d\to 2$, we recover the two dimensional result ($t_c=\varepsilon/2$). The critical exponent ν for correlation length is obtained by

$$\nu = -\frac{1}{\beta'(t_c)} \quad . \tag{41}$$

From (27), ν is $1/(d-2)$ and the conductivity σ becomes

$$\sigma \sim |E - E_c|^s \tag{42}$$

$$s = (d-2)\nu = 1 \quad . \tag{43}$$

The renormalization constant Z is one. Then the exponent η for correlation function at a critical point (mobility edge E_c) and the exponent β for the order parameter become

$$\eta = \zeta(t_c) + 2-d = 2-d \quad , \quad \beta = \frac{\nu}{2}(d-2 + \eta) = 0 \quad . \tag{44}$$

The density of state is interpreted as an order parameter and it does not vanish at a critical point. This is consistent with vanishing exponent β. Our results of the exponents and diffusion constant $D(\omega)$ are derived from the simple form of β-function. The result of β-function is similar to the spherical model which gives the exponent $\nu=1/(d-2)$. The spherical model is a self-consistent approximation for the spin model. For Anderson localization problem, the self-consistent method has been applied [30-33] and the same exponent s and the same ω-dependent diffusion constant have been obtained. However there is a difference between our treatment and self-consistent approximation. The exponent η becomes 2-d in our renormalization group analysis and η becomes zero for the self-consistent method. We have introduced external frequency $-i\omega$ as infrared regulator and we have obtained $q^2 \ln\omega$ term for $\Gamma^{(2)}(q)$. This is well-known infrared divergence for the transverse propagator [34,35]. The symmetrical density and density correlation function (sum of transverse and longitudinal propagator) shows $q^{-2+\eta}$ behavior in $d=2+\varepsilon$ at critical point. At a mobility edge, an interesting q^{-d} behavior [36] is obtained for the density-density correlation function.

4. Symplectic and unitary case

We have discussed the orthogonal, real field variable case. If the impurity has a spin, the system can not be described by the orthogonal matrix since there appear spin-orbit interaction and paramagnetic spin-spin interaction [8,13,14,37] and these interactions are treated in symplectic and unitary Grassmannian nonlinear σ model respectively by the symmetric argument. To confirm this general theory, we have made $1/E_F\tau$ expansion for such cases [12,13] and obtained the coefficients of β-function (27) as $a_1=-1$, $a_2=a_3=0$ for the symplectic case (there is a relation between orthogonal and symplectic case [14,16,17-18]) and $a_1=0$, $a_2=2$, $a_3=0$ for the unitary case (the result of a_2 disagrees with [8]). Here we briefly discuss the diffusion constant D by solving the renormalization group equation (28) in d=2. For the symplectic case (strong spin-orbit interaction), we have ($\beta=t^2$, $\mu=1$)

$$-i\omega = D \exp(\frac{2}{t} - 2D) \tag{45}$$

and for unitary case ($\beta = -2t^3$)

$$-i\omega = D \exp[\frac{1}{2}(D^2 - \frac{1}{t^2})] \quad . \tag{46}$$

5. Lattice nonlinear σ model

We have studied the perturbational treatment for the Anderson localization problem, or equivalently small coupling t-expansion for the nonlinear σ model. Usually, the nonlinear σ model can be studied also in a strong coupling region ($t>>1$) by the formulation on a discrete lattice. The high temperature $1/t$ expansion gives the useful result of the transition. However, the relevant nonlinear σ model is noncompact as we have seen, and the high temperature expansion is difficult. For one dimensional case, we can solve the lattice nonlinear σ model for all region of t [19]. There is no ordered state in $d=1$ and β-function becomes asymptotic free. Although it is not necessary that the β-function of the Anderson localization coincides strictly with the β-function of the lattice nonlinear σ model, the general feature is expected to be similar. For large t ($t=1/E_F\tau \gtrsim 1$), β-functions of Grassmannian nonlinear σ models behave as

$$\beta(t) = -(1+t)\ell n(1+t) \quad . \tag{47}$$

This is same behavior derived by ANDERSON et al. [38] in one dimensional case. For higher dimensions, the study of the lattice model may give additional information to the renormalization group calculation.

Acknowledgements

I wish to thank E. Brézin, S. Kirkpatrick and P.A. Lee for the discussions of a scaling function.

References

1. P.W. Anderson, Phys. Rev. 109, 1492 (1958).
2. D.J. Thouless, in Ill-condensed matter, edited by R. Balian, R. Maynard and G. Toulouse (North-Holland, Amsterdam, 1979) Vol.31.
3. S.F. Edwards, Philos. Mag. 3, 1020 (1958).
4. G. Rickayzen, in The Many Body Problem, Bergen International School of Physics (New York: W.A. Benjamin; 1962).
5. J.S. Langer and T. Neal, Phys. Rev. Lett. 16, 984 (1966).
6. E. Abrahams, P.W. Anderson, D.C. Licciardello and T.V. Ramakrishnan, Phys. Rev. Lett. 42, 673 (1979).
7. L.P. Gorkov, A.I. Larkin and D.E. Khmelnitskii, Sov. Phys. JETP Lett. 30, 228 (1979).
8. D.E. Khmelnitskii, unpublished.
9. R. Oppermann and F.J. Wegner, Z. Phys. B34, 327 (1979).
10. F.J. Wegner, Z. Phys. B35, 207 (1979).
11. E. Brézin, S. Hikami and J. Zinn-Justin, Nucl. Phys. B165, 528 (1980).
12. S. Hikami, Phys. Rev. B24, 2671 (1981).
13. S. Hikami, A.I. Larkin and Y. Nagaoka, Prog. Theor. Phys. 63, 707 (1980).
14. R. Oppermann and K. Jüngling, Phys. Lett. 76A, 449 (1980).
15. K.B. Efetov, A.I. Larkin and D.E. Khmelnitskii, Sov. Phys. JETP 52, 568 (1980).
16. F.J. Wegner, Nucl. Phys. B180, 77 (1981).
17. S. Hikami, Prog. Theor. Phys. 64, 1466 (1980).
18. S. Hikami, Phys. Lett. 98B, 208 (1981).
19. S. Hikami and T. Maskawa, Prog. Theor. Phys. 67, No4. (1982).
20. P.G. de Gennes, Phys. Lett. 38A, 339 (1972).
21. J. des Cloizeaux, J. Phys. (Paris) 36, 281 (1975).

22. S. Hikami, Prog. Theor. Phys. 62, 226 (1979).
23. F.J. Dyson, J. Math. Phys. 3, 140 (1962).
24. M.L. Mehta, Random Matrices and the Statistical Theory of Energy Levels (1967, Academic Press, New York).
25. L.P. Gorkov and G.M. Eliashberg, Sov. Phys. JETP 21, 940 (1965).
26. L. Schäfer and F.J. Wegner, Z. Physik B38, 113 (1980).
27. A. Houghton, J. Jevicki, R.D. Kenway and A.M.M. Pruisken, Phys. Rev. Lett. 45, 394 (1980).
28. A.J. McKane and M. Stone, Ann. Phys. 131, 36 (1981).
29. E. Brézin and J. Zinn-Justin, Phys. Rev. B14, 3110 (1976).
30. D. Vollhardt and P. Wölfle, Phys. Rev. B22, 4666 (1980).
31. A. Kawabata, Solid State Comm. 38, 823 (1981).
32. B. Shapiro, a preprint.
33. Y. Ono, D. Yoshioka and H. Fukuyama, J. Phys. Soc. Japan 50, 2143 (1981).
34. A. McKane and M. Stone, Nucl. Phys. B163, 169 (1980).
35. D.J. Amit and G.B. Kotliar, Nucl. Phys. B170 [FS1], 187 (1980).
36. P.A. Lee, a private communication.
37. P.A. Lee, J. Non-Cryst. Solids 35-36, 21 (1980).
38. P.W. Anderson, D.J. Thouless, E. Abrahams and D.S. Fisher, Phys. Rev. B22, 3519 (1980).

Self-Consistent Diagrammatic Theory of Anderson Localization

P. Wölfle[1,2] and D. Vollhardt[2]

[1]Physik-Department der Technischen Universität München
D-8046 Garching, Fed. Rep. of Germany and

[2]Max-Planck-Institut für Physik und Astrophysik
Werner-Heisenberg-Institut für Physik
D-8000 München, Fed. Rep. of Germany

Abstract

The model of independent electrons interacting with static, randomly distributed, pointlike scatterers is treated in the extended state perturbation theory at zero temperature. We discuss several types of infrared divergent contributions indicating the onset of Anderson localization. Beyond perturbation theory, using a novel classification scheme of diagrams for the total vertex function, a self-consistent equation for the frequency dependent diffusion coefficient is derived. Solving these equations, one finds that for $d \leqslant 2$ states are always localized, while for $d > 2$ particles are delocalized for sufficiently weak disorder. We calculate properties in the critical region, e.g. critical exponents and scaling behavior. Defining a frequency dependent conductance we derive a scaling equation and explicitly calculate the corresponding β-function. By a plausible extension we also investigate the length-dependent conductance and its β-function. Where comparison is possible the results agree with those of renormalization group calculations.

1. Introduction

Over the past three years considerable progress has been made in understanding the problem of localization of a quantum mechanical particle subject to a random potental. In his first formulation of this problem ANDERSON [1] showed the existence of a sharp transition from localized to delocalized behavior, or equivalently, from an insulating to a metallic state, as a function of disorder. Subsequently MOTT [2] introduced the concept of a mobility edge, defined as the critical energy at which the mobility as a function of the Fermi energy was assumed to drop to zero discontinuously.

Recent work has been done along two main lines. First, the analogy to phase transitions has led to the application of modern concepts of the theory of critical phenomena. In particular, the scaling theory by ABRAHAMS et al. [3], based on ideas developed earlier by THOULESS [4], treats the dimensionless conductance g as a function of sample length L as the only relevant scaling parameter. The corresponding β-function is determined for strong disorder by assuming exponential localization and for weak disorder by applying extended state perturbation theory as developed by EDWARDS [5]. By smoothly interpolating between the two limits one finds [3] a continuous conductivity near the transition and, in dimension $d = 2$, weak localization even for arbitrarily small disorder, both being in contradiction to MOTT [2]. Some of these results had been derived earlier by WEGNER [6] applying the real space renormalization group (RG) approach. The assumptions of the scaling theory are justified to some extent by mapping the original Hamiltonian onto field theoretical models like the n-orbitals per site model [7, 8] and the nonlinear σ-models of interacting matrices discussed by WEGNER [9], HIKAMI [10], EFETOV et al. [11] and others.

The second line of approach aims at an approximate description of the full range of physical properties (i.e. not only the critical region), similar in spirit to the van der Waals theory of fluids. Such a theory has first been proposed by GÖTZE [12] in the form of a self-consistent equation for the dynamic density response function. The derivation is based on a mode-coupling formalism previously applied in the theory of liquids by the same author. The theory describes a continuous Anderson transition, with the correct limiting behavior for weak and strong disorder [12] and has been applied to explain the temperature dependence [13,14] and frequency dependence [15] of the conductivity. In the theory [12], however, the conductance does not obey scaling in the sense of ABRAHAMS et al. [3], as is apparent from perturbation theory. Also the critical properties are different from the field theoretical results [9,10]. Given the promising features of this first self-consistent theory of the Anderson transition it seemed worthwhile to try (i) to understand its relation to the field-theoretical approach and (ii) on that basis, to derive a new self-consistent equation which would hopefully describe the critical behavior correctly. By carrying out a systematic analysis of the density response function in terms of diagrams of standard extended state perturbation theory for a model of point scattering centers, we have derived a self-consistent equation for the frequency dependent conductivity [16]. This equation indeed obeys scaling and yields a conductance in detailed agreement with field theoretical results [17]. The same self-consistent equation has subsequently been derived by KAWABATA [18] in a similar way. Modifying the relaxation kernel in the mode-coupling theory [12], PRELOVSEK [19] arrived at the same equation, too. The combined effect of the classical [12] localizaton mechanism and the quantum interference effect [3] has been considered by BELITZ et al. [20].

It is the purpose of this article to review the diagrammatic self-consistent theory [16,17]. In the next section the diagrammatic representation of the density response function will be introduced. The frequency dependent diffusion coefficient $D(\omega)$ is expressed in terms of the irreducible vertex function U. Infrared contributions to U are investigated in some detail in section 3. The most important of these can be expressed in terms of a momentum integral over the density propagator, supplementing the initial derivation of the self-consistent equation on a diagrammatic level. This self-consistent equation for the frequency dependent conductivity and for the length dependent conductance is solved in sections 4 and 5. A brief discussion of the effect of symmetry breaking fields is given in section 6.

2. Density Response Theory

We consider the model of independent electrons interacting with randomly distributed static impurity scattering centers (density n_i), as described by the Hamiltonian

$$H = \sum_{\vec{p},\sigma} \frac{p^2}{2m} a^+_{\vec{p}\sigma} a_{\vec{p}\sigma} + \sum_{\vec{q}} \sum_{\vec{r}_i} V(\vec{q}) e^{i\vec{q}\cdot\vec{r}_i} \rho(\vec{q}) . \quad (1)$$

Here $V(\vec{q})$ is the interaction potential and $\vec{r}_i$ is the position of a single impurity. We are using the second quantized formulation for convenience with field operators $a^+_{\vec{p}\sigma}$, $(a_{\vec{p}\sigma})$ creating (annihilating) electrons in plane wave states of momentum $\vec{p}$ and spin σ in spite of the fact that we are dealing with a single particle problem. In (1) the electron density operator $\rho(\vec{q}) = \sum_{\vec{p},\sigma} a^{+}_{\vec{p}+\vec{q}/2,\sigma} a_{\vec{p}-\vec{q}/2,\sigma}$ has been introduced. The quantity of main interest here is the density response function $\chi(\vec{q},\omega)$ in the thermodynamic limit, averaged over the statistical impurity positions. The limit of zero temperature is considered. The neglect of mutual interactions in our model implies that $\chi(\vec{q},\omega)$ is given by the impurity average of a product of two single particle Green's functions [21,22]

$$\chi(\vec{q},\omega) = \sum_{\vec{p},\vec{p}'} \{ \int_{E_F-\omega}^{E_F} dE\,\phi^{RA}_{\vec{p}\vec{p}'}(E,\omega;\vec{q}) + \int_{0}^{E_F} dE[\phi^{RR}_{\vec{p}\vec{p}'}(E,\omega;\vec{q}) - \phi^{AA}_{\vec{p}\vec{p}'}(E-\omega,\omega;\vec{q})]\} \quad (2a)$$

where

$$\phi^{RA}_{\vec{p}\vec{p}'}(E,\omega;\vec{q}) = -\frac{1}{2\pi i} < G^{R}(\vec{p}_{+},\vec{p}_{+}'; E+\omega)\, G^{A}(\vec{p}_{-}', \vec{p}_{-};E) > , \quad (2b)$$

and $\vec{p}_{+} = \vec{p} + \vec{q}/2$, etc; E_F is the Fermi energy. We will be interested mainly in the macroscopic properties which allow us to restrict our considerations to the regime of small ω,q. Then (2) reduces to

$$\chi(\vec{q},\omega) = \omega\phi(\vec{q},\omega) + N_F + 0\,(\omega \frac{\partial N}{\partial E_F}, q^2) \quad (3)$$

where

$$\phi(\vec{q},\omega) = \sum_{\vec{p},\vec{p}'} \phi^{RA}_{\vec{p}\vec{p}'}(E_F,\omega;\vec{q}) \quad (4)$$

and N_F is the density of states at the Fermi level for both spin components (the density of states for a single spin component will be denoted by $N(E_F)$). In the context of (3) $\phi(\vec{q},\omega)$ may be interpreted as the Kubo density relaxaton function.

The perturbation theory for $\phi^{RA}_{\vec{p}\vec{p}'}(\vec{q}\omega)$ (in the following we shall drop the superscript RA) may be formulated in terms of exact averaged single particle Green's functions [21] $G^{R}_{\vec{p}}(E) = (G^{A}_{\vec{p}}(E))^*$,

$$G^{R}_{\vec{p}}(E) = [E-p^2/2m - \Sigma^{R}_{\vec{p}}(E)]^{-1} \quad (5)$$

where $\Sigma^{R}_{\vec{p}}(E)$ is the self-energy part. In the absence of mutual interactions the self-energy may be shown to be a bounded, nonsingular quantity, in particular varying smoothly as a function of energy near the mobility edge. For the purpose of calculating the leading singular contributions to the density response function it is therefore sufficient to approximate Σ by the lowest order (Born approximation) expression

$$\Sigma^{R}_{\vec{p}}(E_F) \cong -i\gamma , \quad (6)$$

where $\gamma = \pi N(E_F) \int \frac{d\Omega_{p'}}{4\pi} U_0(\vec{p}_F-\vec{p}_F')$ in $d = 3$.

Here $U_o(\vec{p}-\vec{p}') = n_i \mid V(\vec{p}-\vec{p}')\mid^2$ is the bare vertex function. In the following only short-ranged potentials will be considered, where $U_o(\vec{p}-\vec{p}') \cong U_o =$ const.

The perturbation series for $\phi_{\vec{p}\vec{p}'}$ consists of all skeleton diagrams with a parallel particle and hole line (particle-hole propagator), joined by impurity interaction lines in all possible ways. Each particle (hole) propagator represents a Green's function $G^R_{\vec{p}}(E_F+\omega)(G^A_{\vec{p}}(E_F))$, each interaction line a factor U_o. The external momenta are $\vec{p}_+$ and $\vec{p}'_+$ for the particle propagator and $\vec{p}_-$, $\vec{p}'_-$ for the hole propagator, whereas the internal momenta are given by momentum conservation at the vertices.

The following kinetic equation is obeyed [16] by the distribution function $\phi_{\vec{p}}(\vec{q},\omega) = \sum_{\vec{p}'} \phi_{\vec{p}\vec{p}'}(\vec{q},\omega)$

$$[\ \omega-\vec{p}\cdot\vec{q}/m - \Sigma^R_{\vec{p}_+}\ (E_F+\omega) - \Sigma^A_{\vec{p}_-}\ (E_F)\ ]\phi_{\vec{p}}(\vec{q},\omega) = \\ = [G^R_{\vec{p}_+}\ (E_F+\omega) - G^A_{\vec{p}_-}(E_F)]\ [\frac{1}{2\pi i} - \sum_{\vec{p}'} U_{\vec{p}\vec{p}'}\ (\vec{q},\omega)\ \ \phi_{\vec{p}'}(\vec{q},\omega)]\ . \tag{7}$$

The quantity of prime importance in this equation is the irreducible vertex part $U_{\vec{p}\vec{p}'}(\vec{q},\omega)$, defined as the sum of all particle-hole irreducible diagrams. It has been shown that $U_{\vec{p}\vec{p}'}$ obeys the Ward identity [16]

$$\Sigma^R_{\vec{p}_+}\ (E_F+\omega) - \Sigma^A_{\vec{p}_-}(E_F) = \sum_{\vec{p}'} U_{\vec{p}\vec{p}'}(\vec{q},\omega)\ [G^R_{\vec{p}'_+}(E_F+\omega) - G^A_{\vec{p}'_-}(E_F)\ ] \tag{8}$$

which is a consequence of particle number conservation. Summing (7) on $\vec{p}$ taking the identity (8) into account, one finds the continuity equation,

$$\omega\phi(\vec{q},\omega) \ + \ q\phi_j(\vec{q},\omega) \ = \ - N_F + 0\ (\omega \frac{\partial N(E_F)}{\partial E_F}\ ,\ q^2)\ . \tag{9}$$

The current relaxation function

$$\phi_j\ (\vec{q},\omega) = \sum_{\vec{p}}\ (\vec{p}\cdot\vec{q}/m)\ \ \phi_{\vec{p}}(\vec{q},\omega) \tag{10}$$

obeys, for small $\omega,\vec{q}$, the equation [16]

$$[\ \omega + \frac{i}{\tau} K\ (\vec{q},\omega)]\ \phi_j(\vec{q},\omega) \ = \ q\ \frac{v_F^2}{d}\ \phi\ (\vec{q},\omega)\ . \tag{11}$$

The dimensionless current relaxation kernel $K(\vec{q},\omega)$ is defined in terms of the irreducible vertex by

$$K(\vec{q},\omega) = 1 + \frac{\tau}{\pi mn} \sum_{\vec{p},\vec{p}'} (\vec{p}\cdot\hat{q})\ \Delta G_{\vec{p}}\ [U_{\vec{p}\vec{p}'}(\vec{q},\omega) - U^{\circ}_{\vec{p}\vec{p}'}]\ \ \Delta G_{\vec{p}'}(\vec{p}'\cdot\hat{q}) \tag{12}$$

where $\Delta G_{\vec{p}} = G^R_{\vec{p}_+}(E_F+\omega) - G^A_{\vec{p}_-}(E_F)$.

Here n is the particle density and τ is the weak coupling transport relaxation time, given by

$$\frac{1}{\tau} = 2\gamma + \frac{d}{\pi N_F p_F^2} \sum_{\vec{p},\vec{p}'} (\vec{p}\cdot\hat{q})\Delta G_{\vec{p}}\, U^\circ(\vec{p}-\vec{p}')\, \Delta G_{\vec{p}'}(\vec{p}'\cdot\hat{q})$$

$$= 2\pi n_i \sum_{\vec{p}'} \delta(E_F - \frac{p'^2}{2m})\, |V(\vec{p}-\vec{p}')|^2\, (1-\hat{p}\cdot\hat{p}') \quad . \tag{13}$$

From (9) and (11) one obtains for small q,

$$\phi(\vec{q},\omega) = \frac{-N_F}{\omega+iD(\vec{q},\omega)q^2} \tag{14}$$

where the $\vec{q},\omega$ dependent diffusion coefficient is given by

$$D(\vec{q},\omega) = \frac{D_\circ}{K(\vec{q},\omega)} \quad , \tag{15}$$

and $D_\circ = (v_F^2/d)\tau$ is the bare diffusion constant ($v_F = p_F/m$).

The dynamical conductivity $\sigma(\omega)$ and the polarizability $\alpha(\omega)$ are obtained from the q = 0 limit of the diffusion coefficient $D(0,\omega) \equiv D(\omega)$ as

$$\sigma(\omega) = i\omega\alpha(\omega) = e^2 (nd/2E_F)\, D(\omega). \tag{16}$$

In the conducting regime we expect $\sigma(0) > 0$ and hence $\alpha(\omega) \propto (i\omega)^{-1}$, whereas in the insulating regime $\alpha(0) > 0$ and $\sigma(\omega) \propto i\omega$, for small ω, and hence $K(0,\omega) \propto \omega^{-1}$. Within our formalism localization is characterized by the very appearance of a $1/\omega$ -singularity in $K(0,\omega)$.

3. Infrared divergencies of the relaxation kernel K(q,ω)

Infrared divergent terms are quite naturally introduced into the perturbation series of $K(\vec{q},\omega)$ by terms containing singular propagators in the limit $\vec{q},\omega \to 0$. In the model considered the only conserved quantity is the particle number, giving rise to a diffusion pole structure of the density relaxation function at small $\vec{q},\omega$ as exhibited by (14). The integral over the momentum $\vec{q}$ of $\phi(\vec{q},\omega)$ diverges for dimensions $d \leq 2$, even if $D(\omega) > 0$, for $\omega \to 0$, as is the case in the conducting regime:

$$\int dq \frac{q^{d-1}}{\omega+iD(0)q^2} \propto \begin{cases} (D(0)/\omega)^{1/2} & , \quad d=1 \\ \ell n(D(0)/\omega) & , \quad d=2 \end{cases} \quad . \tag{17}$$

Note that a factor of q^2 in the integrand removes this singularity. In the insulating regime, where $D(\omega) \propto -i\omega$, we have a $1/\omega$ singularity from the same integral

$$\int dq \, \frac{q^{d-1}}{\omega+iD(\omega)q^2} \propto \frac{1}{\omega} \int dq \, \frac{q^{d-1}}{1+\alpha(0)(d/2e^2E_F n)\, q^2} \,. \tag{18}$$

Additional factors of q^2 in the integrand do not remove the singularity in the latter case. Obviously contributions of the form (18) may generate a $1/\omega$ singularity in the current relaxation kernel $K(\vec{q},\omega)$.

In lowest order a diffusion pole is generated by summing the particle-hole ladder diagrams shown in Fig.1, yielding

$$\Gamma^0_{\vec{p}\vec{p}'}(\vec{q},\omega) = \frac{2U_0\gamma}{-i\omega+D_0q^2} \,. \tag{19}$$

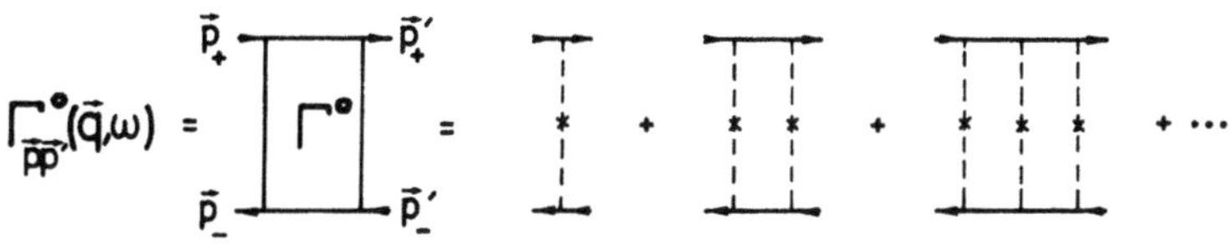

Fig. 1 Particle-hole diffusion propagator $\Gamma^0_{\vec{p}\vec{p}'}(\vec{q},\omega)$

The simplest irreducible diagrams containing Γ^0 are constructed by putting an interaction line across Γ^0 in the four possible ways, as shown in Fig. 2. Substituting $U^D_{\vec{p}\vec{p}'}$, the sum of these four irreducible diagrams, for $U_{\vec{p}\vec{p}'}(\vec{q},\omega)$ into (12) one finds the following contribution to K

$$K^D(\vec{q},\omega) = (\pi nm)^{-1} \sum_{\vec{k}} \frac{q^2+(\vec{k}.\hat{q})^2}{-i\omega+D_0k^2} \,. \tag{20}$$

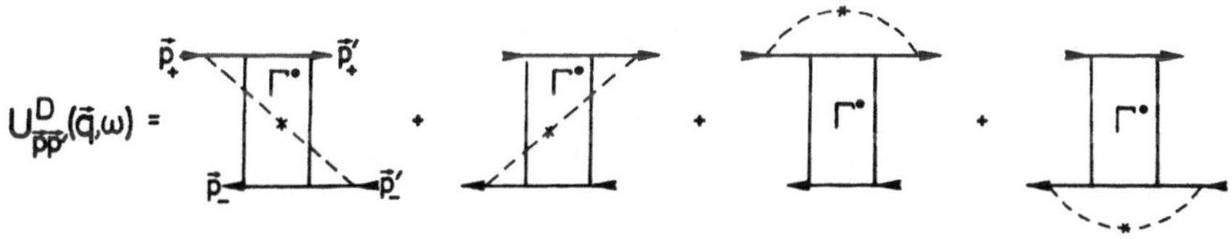

Fig. 2 Simplest irreducible diagrams containing a particle-hole diffusion propagator Γ^0

Whenever K^D is not small compared to unity the diffusion coefficient is changed from its bare value D_0. Accordingly D_0 should be replaced by $D(\vec{q},\omega)$ in the diffusion pole in (20). The resulting self-consistent equation for $D(\vec{q},\omega)$ is essentially identical with the small $\omega,\vec{q}$ limit of the self-consistent equation for $\phi(\vec{q},\omega)$ proposed by GÖTZE [12]. The solution of this equation has been discussed in detail [12-15,23]. Equation (20) has a peculiar q-dependence: in the limit q = 0 the factor k^2 in the integrand removes the explicit infrared divergence for $d \leqslant 2$, but for $q \neq 0$ the divergence is present. This behavior may give rise to the expectation that the singular behavior at $q \neq 0$ is transferred to q = 0 in diagrams containing one diffusion propagator and several or many interaction lines crossing the diffusion propagator. We have shown [16], however, that the infrared divergent contributions are cancelled, i.e. a factor k^2 multiplies the diffusion propagator, within a class of diagrams that is essentially generated by cyclic permutation of the end points of interaction lines on one side of the diffusion propagator. The proof is restricted to momentum independent interaction and an arbitrarily large, but finite number of interaction lines, i.e. diagrams with more than one diffusion propagator are generally excluded.

In addition to the pure diffusion pole singularities, which are caused by particle number conservation and thus have a classical analogue, there exists a second type of singularity which is of pure quantum nature: the particle-particle propagator is singular for small ω, $\vec{Q}$, where $\vec{Q}$ is the total momentum of the two particles [3,24]. This singularity, the so-called Cooper pole or $2k_F$ pole, arises from a subtle quantum interference effect, which only exists if the system is time reversal invariant. It is easily seen that the singularity is exactly the diffusion pole, with a different momentum variable. For a time reversal invariant system the exact energy eigenfunctions $\varphi_n(r)$ may be chosen real, and hence the exact Green's function obeys $G^R(\vec{p},\vec{p}\,',E) = G^R(-\vec{p}\,',-\vec{p},E)$. From this one obtains, together with (2b)

$$\phi_{\vec{p}\vec{p}\,'}(\vec{q},\omega) = \phi_{(\vec{p}-\vec{p}\,'+\vec{q})/2,\ (\vec{p}\,'-\vec{p}+\vec{q})/2}(\vec{p}+\vec{p}\,',\omega)\ . \qquad (21)$$

Equation (21) is demonstrated diagrammatically in Fig. 3.

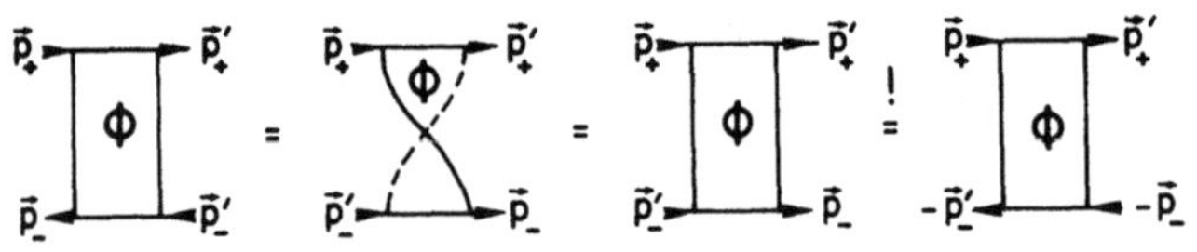

Fig. 3 Diagrammatic demonstration of the crossing symmetry for $\phi_{\vec{p}\vec{p}\,'}(\vec{q},\omega)$ in the case of time reversal invariance; see eq. (21)

The singularity in the particle-hole propagator at $\vec{p}+\vec{p}\,'=0$ is obtained in lowest order by summing the maximally crossed diagrams [24] as shown in Fig. 4, with the result

$$\overset{\circ}{\Lambda}_{\vec{p}\vec{p}\,'}(\vec{q},\omega) = \frac{2\,U_\circ\gamma}{-i\omega+D_\circ(\vec{p}+\vec{p}\,')^2}\ . \qquad (22)$$

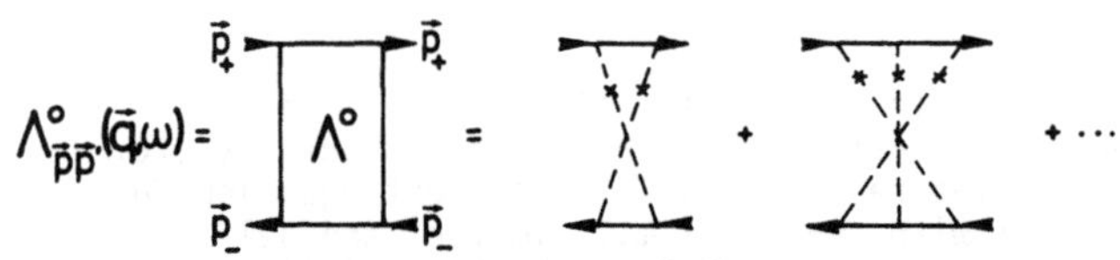

Fig. 4 Maximally crossed diagrams

Any diagram containing $\overset{\circ}{\Lambda}_{\vec{p}\vec{p}'}(\vec{q},\omega)$ contributes mainly for momenta $\vec{p} \simeq -\vec{p}'$, thus connecting the left and right hand side of $\overset{\circ}{\Lambda}$. For instance, substituting $\Lambda^\circ_{\vec{p}\vec{p}'}(\vec{q},\omega)$ in (12), and putting $\vec{p}' = -\vec{p}$ in $\Delta G_{\vec{p}'}(\vec{p}\cdot\vec{q})$, one finds

$$K^{2k_F}(\vec{q},\omega) = [\pi N(E_F)]^{-1} \sum_{\vec{Q}} \frac{1}{-i\omega+D_\circ Q^2} , \qquad (23)$$

where we have introduced $\vec{Q} = \vec{p}+\vec{p}'$. Note that the additional factor of k^2 in (20) originates from the angular integration on $\vec{p}$ and $\vec{p}'$, which requires factors of $\vec{p}.\vec{k}$ and $\vec{p}'\cdot\vec{k}$, by expanding Green's functions $G^R_{\vec{p}+\vec{k}}$, etc, to yield a finite result. The most important diagrams are therefore those, where a Λ° block is crossing diagonally, thus connecting the two external momenta $\vec{p}$ and $\vec{p}'$. We may then classify the diagrams contributing to the total vertex Γ (and therefore to ϕ) in the way depicted in Fig. 5.

In Fig. 5 a wavy line represents Λ°. The vertex Γ is defined as the sum of all diagrams with no interaction line diagonally connecting the end points.

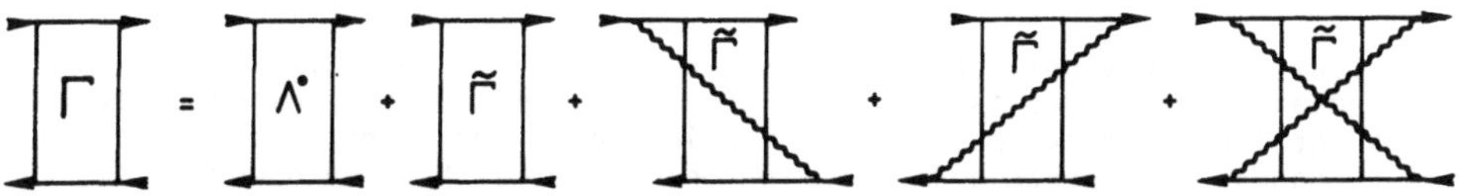

Fig. 5 Exact classification of the total vertex Γ using Cooper-poles (see text)

Observe that only the second diagram of Fig. 5 contains reducible diagrams. Therefore U, the irreducible vertex, is also given by the diagrams of Fig. 5 where, however, $\tilde{\Gamma}$ is replaced by the irreducible part $\tilde{\Gamma}_i$ in the second diagram. The last diagram in Fig. 5 consists of the set of "superdivergent" diagrams [16] containing a Q-integral over $(\Lambda^\circ)^2$ and hence over the diffusion pole squared. Such an irreducible contribution and similar ones giving arbitrary powers of Λ° would give rise to much stronger infrared divergent terms for K and therefore may be expected to be cancelled. The cancellation may be demonstrated by employing the time reversal symmetry relation (21), which may be represented diagrammatically as shown in Fig. 3. By this operation the diagrams of Fig. 5 are untwisted and Λ° is transformed into a sum of particle-hole ladder diagrams. Defining the central part of the last diagram in Fig. 5 as

$$T(\vec{Q},\omega) = \sum_{\vec{p}_1,\vec{p}_2} G^R_{\vec{p}_{1+}} G^A_{\vec{p}_{1-}} \tilde{\Gamma}_{\vec{p}_1\vec{p}_2}(\vec{Q},\omega) \; G^R_{\vec{p}_{2+}} G^A_{\vec{p}_{2-}} , \qquad (24)$$

where $\vec{Q} = \vec{p}+\vec{p}'$ and $\vec{p}_{i\pm} = \vec{p}_i \pm \vec{Q}/2$, it is seen that T consists of all diagrams contributing to the density-relaxation function $\phi(\vec{Q},\omega)$, excluding those with parallel interaction lines (part of a ladder) at the ends. Hence $T(\vec{Q},\omega)$ obeys

$$T = \phi - \phi_\circ - R_\circ\Gamma_\circ T - T\,\Gamma_\circ R_\circ - R_\circ\,\Gamma_\circ\,T\,\Gamma_\circ\,R_\circ \qquad (25)$$

with $R_\circ(\vec{Q},\omega) = \sum_{\vec{p}} G^R_{\vec{p}+} G^A_{\vec{p}-}$ and $(-2\pi i)\,\phi_\circ = (R_\circ/U_\circ)\,\Gamma_\circ = R_\circ/(1-R_\circ U_\circ)$.

Solving (25) for T and substituting the small $\omega,\vec{q}$ expression $R_o = U_o^{-1}[1 - \frac{-i\omega+D_oQ^2}{2\gamma}]$ one finds

$$T(\vec{Q},\omega) = (2\gamma)^{-2}(-i\omega + D_oQ^2)^2 \; [\; \phi(\vec{Q},\omega) - \phi_o(\vec{Q},\omega)\;]. \qquad (26)$$

It is clear that the quadratically divergent term $\Lambda_o^2(\vec{Q},\omega)$ is cancelled.

Substituting (26) and (22) into the diagrams of Fig. 5 one arrives at the following exact expression for the relaxation kernel:

$$K(\vec{q},\omega) = 1 + [2\pi\, N^2(E_F)]^{-1} \sum_{\vec{Q}} \phi(\vec{Q},\omega) + K_1 + K_2\,, \qquad (27)$$

where

$$K_1(\vec{q},\omega) = \text{const.} \sum_{\vec{Q}} \gamma^{-1}(-i\omega+D_oQ^2)\; [\phi(\vec{Q},\omega)-\phi_o(\vec{Q},\omega)]$$

and

$$K_2(\vec{q},\omega) = -i\tau \sum_{\vec{Q}} (\vec{p}\cdot\hat{q})(G^R-G^A)\; \tilde{\Gamma}_i\; (G^R-G^A)\; (\vec{p}\,'\cdot\hat{q})\,.$$

The term K_1 is obviously smaller than the first term and may be safely neglected. The term K_2 is also less divergent, because by construction, it does not contain diagrams in which the left and right vertex are connected by a Cooper pole. Hence all diffusion poles $\Gamma^o(\vec{Q},\omega)$ that may occur in K_2 are multiplied by factors Q^2, originating from terms $(\vec{p}_i\cdot\vec{Q})$ necessary to render finite angular integrals. We may therefore neglect K_2 as well.

The remaining term in (27) is independent of the external momentum $\vec{q}$. More precisely, in deriving (27) we have already neglected terms of the order of $(q\ell)^2$, where $\ell = v_F\tau$ is the mean free path.

4. Self-consistent theory: Dynamical conductivity

Taking the $\vec{Q}$-sum over $\phi(\vec{Q},\omega)$ (the $2k_F$-pole) in (27) as the dominant divergent contribution to $K(\vec{q},\omega)$, one has the following self-consistent equation for $D(\omega)$ and hence the conductivity $\sigma(\omega)$:

$$\frac{D_o}{D(\omega)} = 1 + \frac{k_F^{2-d}}{\pi m} \int_0^{k_o} dk\, \frac{k^{d-1}}{-i\omega + D(\omega)k^2}\;. \qquad (28)$$

It is useful to introduce a dimensionless coupling parameter λ by

$$\lambda = (2\pi\, E_F\, \tau)^{-1},$$

whence $D_o = (\pi d m\lambda)^{-1}$.

In equation (28) a momentum cutoff of the order of the inverse mean free path $k_o \sim 1/\ell = \pi\lambda k_F$ is used. Equation (28) may be written in the following form

$$\frac{D(\omega)}{D_o} = 1 - d\lambda k_F^{2-d} \int_0^{k_o} dk\, \frac{k^{d-1}}{[-i\omega/D(\omega)]+k^2}\;. \qquad (29)$$

In the localized regime we expect $\lim_{\omega\to\infty} [-i\omega/D(\omega)] = \xi^{-2}$, where ξ is the localization length. Indeed the density relaxation function in this regime takes the form [12]

$$\phi(\vec{q},\omega) = \frac{-N_F}{\omega} \frac{1}{1+q^2\xi^2} \tag{30}$$

describing exponential localization with characteristic length ξ. Since in the localized regime $D(\omega=0) = 0$, (29) immediately yields a transcendental equation for ξ,

$$1 = d\lambda\,(k_F\xi)^{2-d} \int_0^{k_0\xi} dy\, \frac{y^{d-1}}{1+y^2}\ . \tag{31}$$

Let us now discuss the solution of (29) and (31). For low dimensions, $d < 2$, the solution of (29) is necessarily of the form $D(\omega) \propto i\omega$, i.e. the particles are localized even for arbitrarily small disorder.

The localization length for $1<d<2$ is obtained from (31) in the limit of small coupling λ as

$$\xi = k_F^{-1}\left[\frac{d}{2}\lambda\ \Gamma\left(\frac{d}{2}\right)\Gamma\left(1-\frac{d}{2}\right)\right]^{\frac{1}{d-2}}\ . \tag{32}$$

For $d = 1$, $k_0\xi \sim 1$ and therefore one obtains $\xi = 4v_F\tau$ which agrees with the exact results [25,26].
For $d = 2$ one finds exponentially weak localization,

$$\xi = k_0^{-1} \exp\left(\frac{1}{2\lambda}\right)\ . \tag{33}$$

The localization length is seen to grow so rapidly with decreasing λ that below $\lambda \sim 0.02$ the localization property does not have practical consequences for laboratory systems.

The static electric polarizability is obtained from the localization length via $\alpha(0) = e^2(\upsilon_0/D_0)\xi^2$.

The leading term of the frequency dependent conductivity is calculated from (29) as

$$\mathrm{Re}\,\sigma(\omega) = \sigma_0\,\omega^2\tau^2\, b_d\ , \tag{34}$$

where

$$b_1 = 512\ , \qquad d = 1$$

$$b_2 = \frac{1}{4\lambda} e^{2/\lambda}\ , \qquad d = 2\ .$$

The level repulsion effect, which leads to a $\omega^2(\ln\omega)^{d+1}$ dependence in $\mathrm{Re}\sigma(\omega)$ is not obtained in the present theory. This may not be surprising as it only shows up in higher order in ω as compared to the leading $\sigma(\omega)\propto -i\omega$ dependence.

For dimensions $d > 2$ the self-consistent equation (29) has a solution $D(\omega=0) \neq 0$ for coupling strength less than a critical value λ_c.

Rewriting (29) in the form

$$\frac{D(\omega)}{D_\circ} = 1 - \frac{d\lambda}{d-2}\left(\frac{k_\circ}{k_F}\right)^{d-2} + d\lambda\left(\frac{Dk_F^2}{-i\omega}\right)^{\frac{2-d}{2}} \int_0^{b(\omega)} dy \frac{y^{d-3}}{1+y^2} \tag{35}$$

where $b(\omega) = [D(\omega)/(-i\omega\tau\, D_\circ)]^{1/2}$, one finds

$$\frac{D(0)}{D_\circ} = 1 - \frac{\lambda}{\lambda_c}, \quad \text{for} \quad \lambda < \lambda_c, \tag{36}$$

where

$$\lambda_c = \frac{d-2}{d}\left(\frac{k_\circ}{k_F}\right)^{2-d}, \quad d \geq 2 \tag{37}$$

or substituting $k_\circ = \pi\lambda k_F$,

$$\lambda_c = \left[\frac{d-2}{d}\,\pi^{2-d}\right]^{1/(d-1)} . \tag{38}$$

The d.c. conductivity $\sigma(0)$ is seen to vanish at the Anderson transition as $\sigma(0) \propto \lambda_c - \lambda$ for any $d > 2$, i.e. the critical exponent of the conductivity is $s = 1$ [17,18,19]. The result (36) implies that

$$\sigma(0) = \sigma_\circ - e^2 \frac{1}{d} c_d k_F^{d-2}, \tag{39}$$

i.e. a disorder independent term is subtracted from the weak coupling conductivity expression $\sigma_\circ = (e^2 n\tau/m)$ ($c_d = \frac{2}{\pi} S_d (2\pi)^{-d}$, with S_d the surface of the d-dimensional unit sphere).

In the localized regime ($\lambda > \lambda_c$) (35) has the general solution for small ω

$$\frac{D_\circ}{D(\omega)} = D_\circ \xi^{-2}/(-i\omega) + O(\omega^\circ) \tag{40}$$

for any dimension d. Expanding the r.h.s. of (35) in powers of ξ^{-1},

$$1 = d\lambda (k_F\xi)^{2-d} \left[\frac{(k_\circ\xi)^{d-2}}{d-2} - \int_0^\infty dy \frac{y^{d-3}}{1+y^2} + O((k_\circ\xi)^{d-4})\right], \quad \text{for } 2 < d < 4$$

and

$$1 = d\lambda (k_F\xi)^{2-d} \left[\frac{(k_\circ\xi)^{d-2}}{d-2} - \frac{(k_\circ\xi)^{d-4}}{d-4} + \int_0^\infty dy \frac{y^{d-5}}{1+y^2} + O((k_\circ\xi)^{d-6})\right] \quad \text{for } d > 4$$

one obtains for $2 < d < 4$ that near the transition ξ diverges as

$$\xi = k_F^{-1} \left[\frac{d-2}{dp_d\lambda} \left|1 - \frac{\lambda}{\lambda_c}\right|\right]^{-1/(d-2)}, \quad 2 < d < 4 \tag{41}$$

where $p_d = \Gamma(\frac{d}{2})\ \Gamma(2-\frac{d}{2})$, i.e. the localization length ξ diverges with the critical exponent $\nu = \frac{1}{d-2}$ [17,19]. For $d > 4$ the critical exponent of the localization length is $\nu = 1/2$ [17,19], independent of dimension, and

$$\xi = k_o^{-1} \left[\frac{d-4}{d-2} \frac{\lambda_c}{\lambda} \quad \left|1 - \frac{\lambda_c}{\lambda}\right| \right]^{-1/2} \quad , \qquad d > 4 \tag{42}$$

At the upper critical dimension $d = 4$ one obtains logarithmic corrections to the classical behavior;

$$\xi = k_o^{-1} \left|1 - \frac{\lambda}{\lambda_c}\right|^{-1/2} \ell n \left|1 - \frac{\lambda}{\lambda_c}\right|^{-1/2} \quad . \tag{43}$$

For $2 < d < 4$ the critical exponents s and ν obey the relation $s=(d-2)\nu$ first derived by WEGNER [6]. The particular values $s=1$ and $\nu = \frac{1}{d-2}$ have been found in ε expansion around $d = 2$ for the nonlinear σ-model of interacting matrices with orthogonal symmetry [9,10,27], which is thought to describe the critical properties of the nonmagnetic impurity Anderson localization problem. By a plausible assumption about the vanishing of all higher order coefficients in the β-function (see below) HIKAMI [28] has recently extended the field-theoretical analysis to the full range $2 < d < 4$. Results for s and ν consistent with ours have also been found by $1/n$-expansion for a n-orbital-per-site model [7,8].

In the conducting regime ($\lambda < \lambda_c$) for $\omega\tau \ll 1-\lambda/\lambda_c$ and in the insulating regime close to the transition ($\lambda \gtrsim \lambda_c$) the upper limit of the k-integral in (35) is $b(\omega) \gg 1$ and may then be taken to infinity. One finds for the frequency dependent conductivity $\sigma(\omega)$

$$\frac{\sigma(\omega)}{\sigma_o} = 1 - \frac{\lambda}{\lambda_c} + p_d \frac{\lambda}{\lambda_c} \left[\frac{-i\omega\tau}{\sigma(\omega)/\sigma_o} \right]^{\frac{d-2}{2}} \tag{44}$$

An essentially identical equation has been derived recently from the non-linear σ-model of orthogonal symmetry [28]. It follows from (44) that $\sigma(\omega)$ has the scaling form [29]

$$\frac{\sigma(\omega)}{\sigma_o} = (-i\omega\tau)^{\frac{d-2}{d}} F_d\,(\omega/\omega_c) \quad , \tag{45}$$

with the limiting values for $\omega \ll \omega_c$

$$\begin{aligned} F\,(\tfrac{\omega}{\omega_c}) &\propto (-i\omega/\omega_c)^{2/d} \quad , & \lambda > \lambda_c \\ F\,(\tfrac{\omega}{\omega_c}) &\propto (-i\omega/\omega_c)^{\frac{2-d}{d}} \quad , & \lambda < \lambda_c \end{aligned} \tag{46}$$

and for $\omega \gg \omega_c$

$$F\,(\tfrac{\omega}{\omega_c}) = \text{const.} \quad , \qquad \lambda \gtrless \lambda_c \ .$$

The latter case ($\omega \gg \omega_c$) describes the scaling region, where $\sigma(\omega) \propto (-i\omega)^{\frac{d-2}{d}}$. The cross-over frequency ω_c scales with the localization length as $\omega_c \propto \xi^{-d}$. Similar scaling forms for $\sigma(\omega)$ have been found by WEGNER [6], GÖTZE [12] (for d = 3), HIKAMI [28] and SHAPIRO and ABRAHAMS [30].

In d = 3 dimensions (44) may be solved explicitly with the result

$$\frac{\sigma(\omega)}{\sigma_0} = \left(\frac{\pi^2}{12}\right)^{1/3} (-i\omega\tau)^{1/3} \left\{\left[\left(1+i\frac{\omega_c}{\omega}\right)^{1/2} - 1\right]^{2/3} + \left[\left(1+i\frac{\omega_c}{\omega}\right)^{1/2} + 1\right]^{2/3} - 2\left(i\frac{\omega_c}{\omega}\right)^{1/3}\right\} , \tag{47}$$

where

$$\omega_c \tau = \frac{4}{9\pi^2} \left(\frac{\lambda}{\lambda_c} - 1\right)^3 . \tag{48}$$

For larger frequencies $\sigma(\omega)$ merges into the well-known Drude behavior $\sigma(\omega)/\sigma_0 = (1-i\omega\tau)^{-1}$.

5. Self-consistent theory: Length-dependent conductance

The scaling properties of the self-consistent theory are most clearly exhibited by the conductance as a function of sample size L. This is also the natural quantity to make contact with the scaling theory [3] and the field-theoretical treatments [6-11,27,28]. Starting from the equation (29) for the frequency dependent diffusion coefficient, we first observe that in a sample of length L (volume L^d) the wave number of any diffusion mode is greater than 1/L. On the other hand, for given disorder the states of the finite sample are localized whenever the states of the infinite system are localized. This is expressed by replacing $-i\omega/D(\omega)$ in the integral of (29) by ξ^{-2}, where ξ is the localization length of the infinite system. The L-dependent diffusion coefficient is then obtained as

$$\frac{D(L)}{D_0} = 1 - d\lambda\xi^2 k_F^{2-d} \int_{1/L}^{k_0} dk \, \frac{k^{d-1}}{1+k^2\xi^2} , \tag{49}$$

or, taking into account the equation (31) for ξ,

$$\frac{D(L)}{D_0} = d\lambda \, (k_F\xi)^{2-d} \int_0^{\xi/L} dy \, \frac{y^{d-1}}{1+y^2} . \tag{50}$$

The calculation of the current density J flowing in a sample of finite length L under the influence of an applied electric field E is nontrivial due to the nonlocal character of the density response of localized electrons. We may determine the electric current by requiring that it be equal and opposite to the diffusion current under open circuit conditions. At the end of the sample (x = L) the current density is then given by $J = eD(L)d(\delta\rho)/dx|_{x=L}$ where $\delta\rho(x)$ is the density change induced by the electric potential $U(x)=eE(L-x)$, given by $\delta\rho(x)=\int_0^L dx' \, \chi(x-x') \, U(x')|_{x=L}$.

Here $\chi(x)$ is the static density response function, obtained by Fourier transforming (30), $\chi(x) = N_F\ [\delta(x)-(2\xi)^{-1}e^{-|x|/\xi}]$. The dimensionless conductance $g(L) = \frac{\hbar}{e^2} L^{d-2} J/E$ is finally obtained as

$$g(L) = c_d\ x^{d-2}\ (1+x)\ e^{-x}\ I_d(x)\ , \tag{51}$$

where $I_d(x) = \int_0^{1/x} dy \frac{y^{d-1}}{1+y^2}$ and we defined $x = L/\xi$.

It is seen that g(L) depends on the coupling constant λ only via the scaled length of the sample, $L/\xi(\lambda)$, and therefore is a scaling function. This is in agreement with the assumptions of the scaling theory of ABRAHAMS et al. [3], as well as the renormalization group treatments of field-theoretical models [6-11,27,28]. The integral in (51) may be performed with the results $I_1(x) = \tan^{-1}(1/x)$, $I_2(x) = (1/2)\ \ln\ [1 + 1/x^2]$ and the relations $I_d(x) = x^{2-d}/(d-2) - I_{d-2}(x)$ for $d > 2$. Thus for d = 1 and d = 2 one finds explicitly

$$\begin{aligned} g(L) &= \frac{2}{\pi^2}\ (\xi/L)\ \tan^{-1}(\xi/L)\ (1 + L/\xi)e^{-L/\xi}, && d = 1 \\ g(L) &= \frac{1}{2\pi^2}\ \ln\ [1 + (\xi/L)^2]\ (1 + L/\xi)e^{-L/\xi}, && d = 2\ . \end{aligned} \tag{52}$$

In the conducting regime (for $d > 2$ and $\lambda < \lambda_c$) we find from (50) by letting $\xi \to 0$

$$g(L) = c_d\ (d-2)^{-1}\ [1 + p_d(L/\xi')^{d-2}], \qquad 2 < d < 4 \tag{53}$$

where the correlation length ξ' is given by (41) for $\lambda < \lambda_c$. Note that g(L) approaches the coupling independent constant $c_d/(d-2)$ for $(L/\xi) \to 0$.

The crucial quantity of the field-theoretical approach is the β-function. It is interesting to calculate $\beta[g(L)]$ from our explicit g(L) and to compare with available results.

From the general result (51) we find

$$\beta\ [g(L)]\ \equiv\ \frac{d\ln g}{d\ln L} = d - 2 - \frac{c_d}{g} \frac{(1 + x)e^{-x}}{1+x^2} - \frac{x^2}{1+x}\ , \tag{54}$$

where $x = L(g)/\xi$ and L(g) is the inverse function of g(L).

The exponential dependence of g(L) for large L/ξ leads to

$$\beta(g) = \ln g + 0\ [\ln\ (\ln \frac{1}{g})]\ , \quad g \ll 1. \tag{55}$$

for all dimensions. On the other hand, for large g and correspondingly small x one finds

$$\beta = d - 2 - \frac{c_d}{g} - x^2\ (1 - \frac{3}{2}\frac{c_d}{g}) + 0\ (x^3)\ . \tag{56}$$

For $d \leqslant 2$ we find $\beta(g) < 0$ for any finite g, implying localization. For $d = 1$ one calculates $x = (\pi g)^{-1}$ and hence β apparently may be expanded in a power series in 1/g. However, for $d = 2$ one finds $x = e^{-\pi^2 g}$ and hence the correction terms to the leading term $-\frac{c_d}{g}$ are nonanalytic for $\frac{1}{g} \to 0$. This is consistent with the finding of field-theoretical calculations [9,10,27] that the coefficients of g^{-n} for $n = 2,3,4$ are zero, and the conjecture that all higher coefficients are zero, too [28].

For $d > 2$ the β-function crosses zero at $g = g_c = c_d/(d-2)$. In the regime $g > g_c$, where g scales as L^{d-2} for large L, β is given by (56) with $x = 0$, since the localization length is infinite in the conducting regime. The slope of β at g_c is continuous and equal to

$$\frac{d\beta}{d\ell ng}\bigg|_{g_c} = \frac{1}{\nu} \quad , \qquad d > 2$$

as expected from the scaling theory [3], where $\nu = (d-2)^{-1}$ for $2 < d < 4$ and $\nu = 1/2$ for $d > 4$.

In Fig. 6 we have plotted the scaling function $\beta(g)$ versus ℓng for dimensions d = 1, 2, 3, 4.

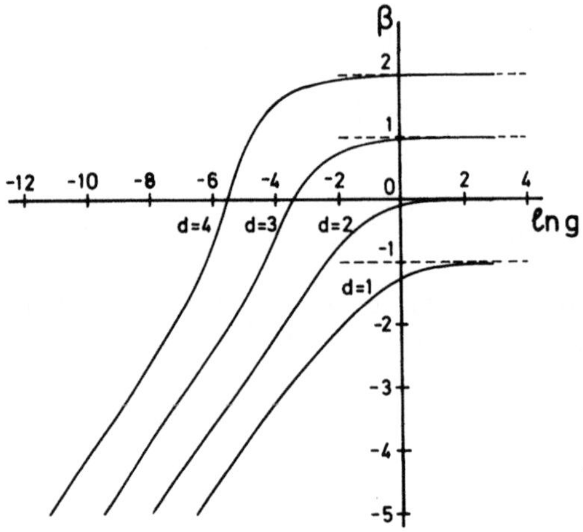

Fig. 6 The β-function for the length dependent conductance g(L), $\beta[g(L)] = d\ell ng/d\ell nL$, for dimensions d=1,2,3,4

One may define a frequency dependent conductance $g(\omega)$ in a volume ξ^d - in analogy to g(L) - by

$$g(\omega) = (e^2/\hbar)^{-1}\,\sigma(\omega)\xi^{d-2} \,. \tag{57}$$

It follows from (44), however, that $g(\omega)$ is not a proper scaling parameter because it goes as $\omega^{(d-2)/d}$ for $\omega \gg \omega_c$, i.e. in the scaling regime. Therefore one has to introduce a modified scaling parameter $G(\omega) = (-i\omega/\omega_c)^{(2-d)/d}\, g(\omega)$ so that for $\omega \gg \omega_c$ $G(\omega)$ is constant. Eq. (44) then takes the form

$$\frac{d-2}{c_d p_d}\, G = \pm\, z^{(2-d)/d} + G^{(2-d)/2}, \; d < 4 \tag{58}$$

where $z = -i\omega/\omega_c$ and

$$\omega_c = \frac{e^2}{\hbar} \frac{D_0}{\sigma_0} \xi^{-d} \tag{59}$$

and the upper (lower) sign corresponds to the conducting (insulating) regime. It is obvious that G is a scaling function with respect to frequency. The corresponding β-function can easily be calculated. For $d < 4$ one obtains

$$\beta[g(\omega)] = \frac{d\ell nG}{d\ell n\omega} = \frac{2}{d} \frac{(2-d)G^{d/2}+c_d p_d}{2G^{d/2}+c_d p_d} . \tag{60}$$

Observe, that for $d \leqslant 2$ the β-function does not change sign, corresponding to the absence of an Anderson transition, while for $d > 2$ β goes through zero at $G_c = [c_d p_d/(d-2)]^{2/d}$. Also note, that for $d = 2$ $\beta[G(\omega)]=(1+2\pi^2 G)^{-1}$ is an analytic function of G in contrast to $\beta[G(L)]$.

6. Conclusion

In this article we have presented a self-consistent theory of the Anderson localization problem with particular emphasis on the critical regime. We find a lower critical dimensionality $d = 2$ and an Anderson transition for $d > 2$, with the static conductivity vanishing continuously as a linear function of the disorder parameter. The localization length ξ diverges at the transitions with the critical exponent $\nu= 1/(d-2)$ for $2 < d < 4$ and $\nu = 1/2$ for $d > 4$. The dimensionless frequency dependent conductance (in volume ξ^d) as well as the length dependent conductance (in volume L^d) vary with disorder only via a scaled frequency ω/ω_c (where $\omega_c \propto \xi^{-d}$) and a scaled length L/ξ. The scaling functions are given explicitly. The corresponding β-functions agree with renormalization group calculations where comparison is possible. For $\omega \gg \omega_c$ (and $\omega\tau \lessapprox 1$), i.e. in the scaling regime, and $d > 2$, the conductivity in the localized and extended regimes varies as $\sigma(\omega) \propto (-i\omega)^{(d-2)/d}$. For $\omega \ll \omega_c$ in the localized regime, the real part of the conductivity is obtained as $\mathrm{Re}\sigma(\omega) \propto \omega^2$. This is in disagreement with the generally accepted result $\mathrm{Re}\sigma(\omega) \propto \omega^2[\ell n\omega]^{d+1}$, where the logarithm is due to a level repulsion effect for pairs of impurities [31]. Such an effect might be included by formulating the theory in terms of localized states.

The q-dependence of the diffusion coefficient is not understood yet within the present theory: whereas from scaling arguments [3,32] one expects $D(\vec{q},\omega) \propto q^{d-2}$ for $q\xi \gg 1$, we have not found any q-dependence on this scale from the diagrammatic calculation of $D^{-1}(\vec{q},\omega)$ in terms of the irreducible vertex function. Also the behavior at still larger wave-vectors and frequencies has not been considered. The cutoff in the wave-vector at the inverse mean free path, which is sufficient in the critical regime, should be replaced by a more realistic description of the density relaxation function $\phi(\vec{q},\omega)$ for larger q and ω values, as e.g. in [12].

A problem of considerable interest is the influence of symmetry-breaking fields on the Anderson transition. The quantum scattering interference effect causing the "Cooper"-pole contribution to the diffusion coefficient, which is ultimately responsible for localization in the theory presented here, is destroyed by time reversal invariance breaking fields, like a magnetic field [33,34] or magnetic impurities [34,35]. In this event, localization can only be caused by particle-hole diffusion pole divergencies. For the case of magnetic impurities we are able to show that infrared divergencies in K from a single diffusion pole are again cancelled in the way discussed in section 3. Such contributions from diagrams containing two and more diffusion poles have been claimed to cancel by KHMELNITSKII [36] and EFETOV et al. [11]. This would imply the existence of an Anderson transition in d = 2 and d = 1. This is in disagreement with results by HIKAMI [19,28] and OPPERMANN [37], who find a lower critical dimension of d = 2, implying that infrared divergent terms exist in $d \leqslant 2$. Given the leading infrared divergent contribution to $K(\vec{q},\omega)$, if it exists, one might again try to derive a self-consistent equation. In the case of an external magnetic field, a self-consistent theory along the lines of the present work has been proposed [38].

Finally, the effect of Coulomb interactions between the particles is expected to be important for a metal-insulator transition. It is hoped that a self-consistent theory of this more complicated problem may be developed on the basis of the present work.

References

[1] P.W. Anderson, Phys.Rev. 109, 1492 (1958).
[2] N.F. Mott, Philos.Mag. 13, 989 (1966); Adv. Phys. 16, 49 (1967).
[3] E. Abrahams, P.W. Anderson, D.C. Licciardello, T.V. Ramakrishnan, Phys.Rev.Lett. 42, 673 (1979).
[4] D.J. Thouless, Phys.Rep. 13C, 93 (1974).
[5] S.F. Edwards, Philos.Mag. 8, 1020 (1958).
[6] F.J. Wegner, Z.Phys. B25, 327 (1976).
[7] F.J. Wegner, Phys.Rev. B19, 783 (1979).
[8] R. Oppermann, F. Wegner, Z.Phys. B34, 327 (1979).
[9] F.J. Wegner, Z.Phys. B35, 207 (1979).
[10] S. Hikami, Phys.Rev. B24, 2671 (1981).
[11] K.B. Efetov, A.I. Larkin, D.E. Khmelnitskii, JETP (Soviet) 52, 568 (1980).
[12] W. Götze, Solid State Commun. 27, 1393 (1978); J.Phys. C12, 1279 (1979); Philos.Mag. 43, 219 (1981).
[13] A. Gold, W. Götze, J.Phys. C14, 4049 (1981).
[14] D. Belitz, W. Götze, Philos.Mag. B43, 517 (1981).
[15] A. Gold, S.J. Allen, B.A. Wilson, D.C. Tsui, Phys.Rev. B (in press).
[16] D. Vollhardt, P. Wölfle, Phys.Rev.Lett. 45, 842 (1980); Phys.Rev. B22, 4666 (1980).
[17] D. Vollhardt, P. Wölfle, to be published.
[18] A. Kawabata, Solid State Commun. 38, 823 (1981).
[19] P. Prelovšek, in *Recent Developments in Condensed Matter Physics*, ed. by J.T. Devreese (Plenum Press 1981).
[20] D. Belitz, A. Gold, W. Götze, Z.Phys. B44, 273 (1981).
[21] A.A. Abrikosov, L.P. Gor'kov, I.E. Dzyaloshinskii, *Methods of Quantum Field Theory in Statistical Physics*, (Prentice Hall, Eaglewood Cliffs, 1969).
[22] S.V. Maleev, B.P. Toperverg, JETP (Soviet) 42, 734 (1976).
[23] W. Götze, P. Prelovšek, P. Wölfle, Solid State Commun. 30, 369 (1979).

[24] L.P. Gor'kov, A.I. Larkin and D.E. Khmel'nitskii, JETP (Soviet) Lett. 30, 228 (1979).
[25] V.L. Berezinsky, JETP (Soviet) 38, 620 (1974).
[26] A.A. Abrikosov, I.A. Ryshkin, Adv.Phys. 27, 147 (1978).
[27] A.J. McKane, M. Stone, Ann.Phys. 131, 131 (1981).
[28] S. Hikami, these Proceedings.
[29] B. Shapiro, Phys.Rev. B (in press).
[30] B. Shapiro, E. Abrahams, Phys.Rev. B24, 4889 (1981).
[31] N.F. Mott, Philos. Mag. 22, 7 (1970).
[32] Y. Imry, J.Appl.Phys. 55, 1812 (1981).
[33] B.L. Altshuler, D.E. Khmelnitskii, A.I. Larkin, P.A. Lee, Phys.Rev. B22, 5142 (1980).
[34] S. Hikami, A.I. Larkin, Y. Nagaoka, Progr.Theor.Phys. 63, 707 (1980).
[35] P.A. Lee, J.Non.Cryst.Sol. 35, 21 (1980).
[36] D.E. Khmelnitskii, Nordita preprint (1980).
[37] R. Oppermann, J.Phys. C14, 3757 (1981).
[38] Y. Ono, D.Yoshioka, H. Fukuyama, J.Phys.Soc. Japan 50, 2143 (1981); D. Yoshioka, Y. Ono, H. Fukuyama, J.Phys.Soc. Japan 50, 3419 (1981); D. Yoshioka, these Proceedings.

Self-Consistent Treatment of Two-Dimensional Anderson Localization in the Absence of Time Reversal Symmetry

D. Yoshioka

The Institute for Solid State Physics, The University of Tokyo
Roppongi, Minato-ku, Tokyo 106, Japan

A self-consistent scheme to calculate the dynamical conductivity $\sigma(\omega)$, which is developed by VOLLHARDT and WÖLFLE, is refined and applied to systems without the time reversal symmetry. The conductivity is calculated in the presence of weak magnetic field, spin-orbit scattering or paramagnetic impurities for two-dimensional systems with $E_F\tau >> 1$, where E_F is the Fermi energy and τ is the lifetime of the electron. It is shown that the weak magnetic field does not destroy the localization, whereas the spin-orbit scattering or the spin-flip scattering by paramagnetic impurities makes the static conductivity $\sigma(0)$ finite.

1. Introduction

ABRAHAMS et al.[1] have shown by the scaling theory that the two-dimensional electrons in static random potential are always localized. This work encouraged diagrammatic perturbational approach to the problem of the Anderson localization, since the localization occurs even for the system where the impurity scattering is weak and perturbational calculation is valid. The simplest model system for such an investigation is mutually non-interacting electrons scattered by short-range impurity potentials with the condition that $E_F\tau >> 1$, where E_F is the Fermi energy and τ is the lifetime of electrons at the Fermi level.

The early investigation on such a system revealed that the so-called mostly crossed diagrams play an important role in the calculation of the dynamical conductivity $\sigma(\omega)$ in the weakly localized regime, $\omega\tau=O(1)$.[2,3] The mostly crossed diagrams are important, because they have a diffusion pole due to the time reversal symmetry. VOLLHARDT and WÖLFLE [4,5] developed a theory in which the diffusion constant is renormalized self-consistently. They obtained the dynamical conductivity which behaves reasonably even in the strongly localized regime, $\omega\tau \simeq 0$, and confirmed the scaling theory of [1] diagrammatically.

Now we consider a little more complicated systems, namely the systems where in addition to the normal impurities a weak magnetic field, spin-orbit scattering, or paramagnetic impurities are present. These systems have already been investigated in the weakly localized regime [6,7], and it has been shown that the behavior of $\sigma(\omega)$ differs from that for the system with the normal impurities only. Such difference comes from the lack of the time reversal symmetry in these systems. Since the theory of [4] is applicable only for systems with the time reversal symmetry, it cannot be applied in the present case. We have extended it to be applicable to the systems without the symmetry, and have calculated $\sigma(\omega)$ in these systems [8,9,10]. (The spin-orbit scattering does not destroy the symmetry. However the spin dependent scattering amplitude requires the extension of the theory in [4] just as for the system with paramagnetic impurities.) In this paper we explain our method and show the result of calculations. We show that the spin-orbit scattering or spin-flip scattering due to the paramagnetic impurities destroy the localization.

Our model is presented in section 2. In section 3 we explain our method by applying it to a system with normal impurities only. In section 4 we apply the method to the system where weak magnetic field, spin-orbit scattering or paramagnetic impurities are present. Since the application of our method to the systems without the time reversal symmetry is straightforward, we only indicate the necessary modifications. The results of the calculation are given for each perturbation. Finally discussion is given in section 5.

2. Model systems

We consider two-dimensional electrons in short-range normal impurity potential, short-range paramagnetic impurity potential and weak magnetic fields. We take the co-ordinate system such that the electrons move in the xy-plane at z=0. The effect of impurity scattering is treated within the Born approximation. The scattering amplitude depends on the direction of the momenta and spin states of the scattered particles. Later we need two types of scattering amplitude $\Gamma^{(0)\alpha\beta\gamma\delta}_{pp'}(q)$ and $\Lambda^{(0)\alpha\beta\gamma\delta}_{pp'}(q)$, which are shown diagramatically in Fig.1. Here $|p|\simeq|p'|\simeq p_F$. $|q|\ll p_F$, where p_F is the Fermi momentum, $p_F^2/2m = E_F$, and α,β,γ and δ denote the spin state of the particles. The amplitudes are given by

$$\Gamma^{(0)\alpha\beta\gamma\delta}_{pp'}(q) = \frac{1}{2\pi N(E_F)}\left[\frac{1}{\tau_0}\delta_{\alpha,\beta}\delta_{\gamma,\delta} + \left(\frac{1}{\tau_{s,i}} + \frac{1}{\tau_{so,i}}\right)\sigma^i_{\alpha\beta}\sigma^i_{\gamma\delta}\right] \quad , \tag{2.1}$$

and

$$\Lambda^{(0)\alpha\beta\gamma\delta}_{pp'}(q) = \frac{1}{2\pi N(E_F)}\left[\frac{1}{\tau_0}\delta_{\alpha,\beta}\delta_{\gamma,\delta} + \left(\frac{1}{\tau_{s,i}} - \frac{1}{\tau_{so,i}}\right)\sigma^i_{\alpha\beta}\sigma^i_{\gamma\delta}\right] \quad . \tag{2.2}$$

In these equations σ is the Pauli matrix and the relaxation times are given by

$$\tau_0^{-1} = 2\pi n_i V^2 N(E_F) \quad , \tag{2.3}$$

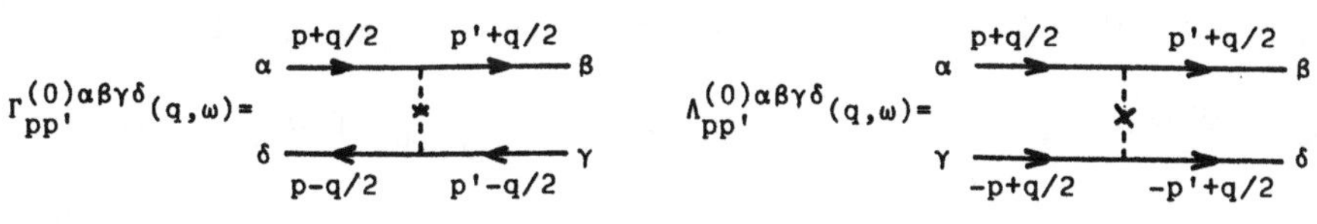

<u>Fig.1</u> The scattering amplitudes $\Gamma^{(0)\alpha\beta\gamma\delta}_{pp'}(q,\omega)$ and $\Lambda^{(0)\alpha\beta\gamma\delta}_{pp'}(q,\omega)$

$$\tau_{so,z}^{-1} = 2\pi n_i V^2 \frac{1}{4m^2c^2} \overline{(p\times p')_z^2}\, N(E_F) \quad , \tag{2.4}$$

$$\tau_{s,z}^{-1} = 2\pi n_s I^2 \langle S_z^2 \rangle \quad , \tag{2.5}$$

$$\tau_{s,x}^{-1} = \tau_{s,y}^{-1} = 2\pi n_s I^2 \langle S_x^2 \rangle \quad , \tag{2.6}$$

where $\overline{(p\times p')_z^2}$ means angular average, $\langle S_i^2 \rangle$ is impurity average of the expectation value of S_i^2, n_i and n_s are the concentration of the impurities, and $N(E_F)$ is the density of states at the Fermi level. Since we are considering a strictly two-dimensional system, the z-component of the momentum p is zero, hence $\tau_{so,x}^{-1}=\tau_{so,y}^{-1}=0$. Namely there is no spin-flip scattering by the spin-orbit interaction. We have assumed isotropy in the xy-plane in the spin space, hence $\tau_{s,x}=\tau_{s,y}$. Due to this condition the total spin is conserved in the spin-flip scattering. We consider a system where $\tau_{s,i}, \tau_{so,z} \gg \tau_0 \gg E_F^{-1}$.

The magnetic field is applied parallel to the z-axis. We consider only weak magnetic field so that we can treat the effect of the field quasi-classically. Namely we replace the momentum operator p by quasimomentum operator $\pi=p+(e/c)A$ where A is the vector potential. We also consider the spin Zeeman energy $1/2g\mu_B H\sigma_z$, where g is the g-factor, and μ_B is the Bohr magneton.

3. Conductivity of the normal system

In this section we consider a normal system, namely H=0 and $\tau_{s,i}^{-1}=\tau_{so,i}^{-1}=0$. Since there is no mixing of the spin states, we suppress the spin indices.

The conductivity at T=0 is given by the Kubo formula as follows

$$\sigma(\omega) = \frac{e^2}{\pi m^2} \sum_p \sum_{p'} p_x p'_x \langle G^R(p,p';E_F+\omega) G^A(p',p;E_F)\rangle \quad , \tag{3.1}$$

where the bracket means the impurity average, and $G^R(p,p';E_F)$ $(G^A(p,p';E_F))$ is retarded (advanced) Green function. We follow the theory of VOLLHARDT and WÖLFLE [4], and express $\sigma(\omega)$ in terms of the current relaxation kernel $M_1(q,\omega)$. We first consider the following function,

$$\phi_{pp'}(q,\omega) = -\frac{1}{2\pi i} \langle G^R(p_+,p'_+;E_F+\omega) G^A(p'_-,p_-;E_F)\rangle \quad , \tag{3.2}$$

where $p_\pm=p\pm q/2$. This $\phi_{pp'}(q,\omega)$ satisfies the Bethe-Salpeter equation:

$$\phi_{pp'}(q,\omega) = G^R(p_+;E_F+\omega) G^A(p_-;E_F) \left[-\frac{1}{2\pi i}\delta_{p,p'} + \sum_{p''} U_{pp''}(q,\omega)\phi_{p''p'}(q,\omega)\right] , \tag{3.3}$$

where

$$G^{R,A}(p;\omega)\delta_{p,p'} = \langle G^{R,A}(p,p';\omega)\rangle = \delta_{p,p'} \left[\omega - \frac{p^2}{2m} - \Sigma^{R,A}(p;\omega)\right]^{-1} \quad , \tag{3.4}$$

and $U_{pp'}(q,\omega)$ is the vertex function that is irreducible with respect to the particle-hole channel. Eq.(3.3) is rewritten into a kinetic equation:

$$[\omega - p\cdot q/m - \Sigma^R(p_+;E_F+\omega) + \Sigma^A(p_-;E_F)]\phi_{pp'}(q,\omega)$$
$$= \Delta_+G_p[\frac{1}{2\pi i}\delta_{p,p'} - \sum_{p''} U_{p,p''}(q,\omega)\phi_{p''p'}(q,\omega)] \quad , \qquad (3.5)$$

where $\Delta_+G_p = G(p_+;E_F+\omega) - G(p_-;E_F)$. We multiply this equation by 1, $p\cdot q/m$ and $p'\cdot q/m$, respectively, and sum over p and p', then we obtain equations for $\phi(q,\omega) = \Sigma_{p,p'}\phi_{pp'}(q,\omega)$, $\phi_j(q,\omega) = \Sigma_{pp'}(p\cdot q/m)\phi_{pp'}(q,\omega)$, and $\phi_{jj}(q,\omega) = \Sigma_{pp'}(p\cdot q/m)(p'\cdot q/m)\phi_{pp'}(q,\omega)$:

$$\omega\phi(q,\omega) - \phi_j(q,\omega) = -N(E_F) \quad , \qquad (3.6)$$

$$[\omega + M_1(q,\omega)]\phi_j(q,\omega) - q^2(E_F/m)\phi(q,\omega) = 0 \quad , \qquad (3.7)$$

$$\omega\phi_j(q,\omega) - \phi_{jj}(q,\omega) = 0 \quad , \qquad (3.8)$$

$$M_1(q,\omega) = 2i\gamma + \frac{i}{\pi N(E_F)p_F^2q^2}\sum_p\sum_{p'}(p\cdot q)\Delta_+G_pU_{pp'}(q,\omega)\Delta_+G_{p'}(p'\cdot q) \quad , \qquad (3.9)$$

where $2\gamma = 1/\tau_0 = i[\Sigma^R(p_F;E_F) - \Sigma^A(p_F;E_F)]$. In deriving (3.6) and (3.8) we have used the Ward identity:

$$\Sigma^R(p_+;E_F+\omega) - \Sigma^A(p_-;E_F) = \sum_{p'} U_{pp'}(q,\omega)\Delta_+G_{p'} \quad , \qquad (3.10)$$

and in deriving (3.7) we have used the expansion of $\Sigma_{p'}\phi_{pp'}(q,\omega)$ in angular variables of p:

$$\sum_{p'}\phi_{pp'}(q,\omega) \simeq -[2\pi i N(E_F)]^{-1}\Delta_+G_p\sum_{p'}\sum_{p''}[1+2(p\cdot q)(p''\cdot q)/p_F^2q^2]\phi_{p''p'}(q,\omega). \qquad (3.11)$$

From (3.6), (3.7) and (3.8) we obtain the conductivity in terms of M_1:

$$\sigma(\omega) = \frac{e^2}{\pi m^2}\phi_{jj}(0,\omega) = i\frac{ne^2}{m}\frac{1}{\omega + M_1(0,\omega)} \quad . \qquad (3.12)$$

The most dominant contribution to $M_1(0,\omega)$ is obtained when we use for $U_{pp'}(q,\omega)$ the diagram shown in Fig.2(a), where $\psi_{pp'}(q,\omega)$ is defined as

$$\psi_{pp'}(q,\omega) = -\frac{1}{2\pi i}\langle G^R(p_+,p_+';E_F+\omega)\, G^A(-p_-,-p_-';E_F)\rangle \quad , \qquad (3.13)$$

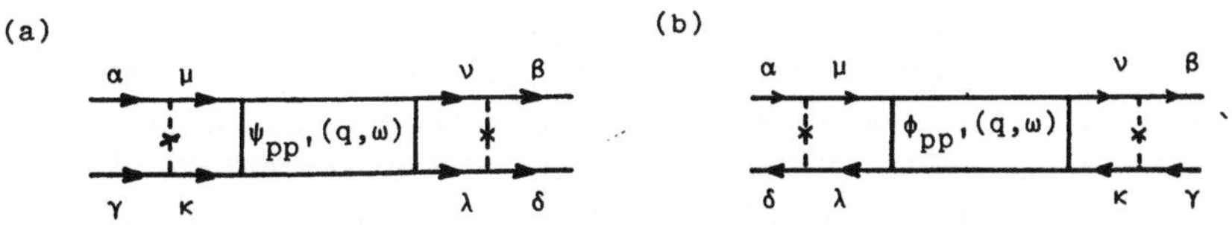

Fig.2 The most inportant diagrams for $U_{pp'}^{\alpha\beta\gamma\delta}(q,\omega)$, (a), and for $V_{pp'}^{\alpha\beta\gamma\delta}(q,\omega)$, (b)

This is because $\psi(q,\omega)\equiv\Sigma_{p,p'}\psi_{pp'}(q,\omega)$ is equal to $\phi(q,\omega)$ when there is the time reversal symmetry and has a diffusion pole for small q and ω. VOLLHARDT and WÖLFLE used this symmetry; they substituted the expression of $\phi(q,\omega)$ for $\psi(q,\omega)$ and obtained a self-consistent equation for $M_1(0,\omega)$. However the diffusion pole of $\psi(q,\omega)$ is suppressed in the systems without the time reversal symmetry, and $\psi(q,\omega)$ becomes different from $\phi(q,\omega)$. Therefore it is necessary for us to obtain an expression for $\psi(q,\omega)$. Although the suppression of the diffusion pole makes the contribution of $\psi(q,\omega)$ to $M_1(q,\omega)$ smaller, it still gives the most dominant contribution to $M_1(q,\omega)$ if the perturbation which destroys the symmetry is weak. Hence we will use the same diagram for $U_{pp'}(q,\omega)$ for such systems, too. The equation for $\psi_{pp'}(q,\omega)$ has a similar form to that for $\phi_{pp'}(q,\omega)$, and $\psi(q,\omega)$ is given by

$$\psi(q,\omega)=\frac{-N(E_F)}{\omega-2\gamma D_0q^2/M_2(q,\omega)} \quad , \tag{3.14}$$

$$M_2(q,\omega) = 2i\gamma + \frac{i}{\pi N(E_F)p_F^2q^2}\sum(p\cdot q)\Delta_-G_p V_{pp'}(q,\omega)\ \Delta_-G_{p'}(p'\cdot q), \tag{3.15}$$

$$\Delta_-G_p = G^R(p_+;E_F+\omega) - G^A(-p_-;E_F) \quad , \tag{3.16}$$

where $D_0=E_F/2m\gamma$, and $V_{pp'}(q,\omega)$ is the vertex function that is irreducible with respect to the particle-particle channel. For $M_2(q,\omega)$ the most dominant contribution is obtained when the diagram shown in Fig.2(b) is used for $V_{pp'}(q,\omega)$. Hence we get self-consistent equations for $M_1(0,\omega)$ and $M_2(0,\omega)$:

$$M_1(0,\omega) = 2i\gamma - 2U_0\sum_q[\omega-2\gamma D_0q^2/M_2(0,\omega)]^{-1} \quad , \tag{3.17}$$

$$M_2(0,\omega) = 2i\gamma - 2U_0\sum_q[\omega-2\gamma D_0q^2/M_1(0,\omega)]^{-1} \quad . \tag{3.18}$$

In the present system there is the time reversal symmetry and M_1 and M_2 are the same. However they are not the same when the symmetry is broken.

4. Conductivity of the system without the time-reversal symmetry

4.1 Weak magnetic field

We use quasiclassical approximation for the magnetic field. We replace the momentum p by p+eA/c. Since the momentum q in $\phi_{pp'}(q,\omega)$ is a total momentum of particle-hole pair which has no net charge, it suffers no change by the magnetic field. On the other hand q in $\psi_{pp'}(q,\omega)$ is a total momentum of particle-particle pair, which has total charge 2e, so that it is replaced by (q+2eA/c). Hence the self-consistent equations for $M_1(0,\omega)$ and $M_2(0,\omega)$ are given by

$$M_1(0,\omega) = 2i\gamma - 2U_0 \sum_k [\omega - 2\gamma D_0(k+\frac{2e}{c}A)^2 / M_2(0,\omega)]^{-1} \quad , \tag{4.1}$$

$$M_2(0,\omega) = 2i\gamma - 2U_0 \sum_q [\omega - 2\gamma D_0\, q^2/M_1(0,\omega)]^{-1} \quad . \tag{4.2}$$

The summation over k in (4.1) is done by replacing the operator $(k+2eA/c)^2$ with its eigenvalue $4m\omega_c(N+1/2)$ and by summing over N. The upper limit of the summations in (4.1) and (4.2) are cut off at N_c or q_c such that $4mD_0\omega_c(N_c+1/2)\tau=1$ or $D_0k_c^2\tau=1$. The equations are rewritten by $\Phi(\omega)=-i\tau M_1(0,\omega)$ and $\Psi(\omega)= -i\tau M_2(0,\omega)$ as

$$\Phi(\omega) = 1 - \lambda\Psi(\omega)\ \{\ell n \frac{2\omega_c\tau}{\pi\lambda} + \psi[\frac{1}{2} - i\frac{\pi\lambda}{2}\frac{\omega}{\omega_c}\Psi(\omega)]\} \quad , \tag{4.3}$$

$$\Psi(\omega) = 1 - \lambda\Phi(\omega)\ \ell n[-i\Phi(\omega)\omega\tau] \quad , \tag{4.4}$$

where ψ is the digamma function, and $\lambda=(2\pi E_F\tau)^{-1}$.

At small ω and H we expand $\Phi(\omega)$ in the power series of ω and H. Then the conductivity is given by

$$\frac{\sigma(\omega;H)}{\sigma(\omega;0)} = 1 + \frac{(\omega_c\tau)^2}{6\pi^2\lambda^2(2-\lambda)} e^{2/\lambda} + \frac{\pi^2}{4}(g\omega_c\tau)^2 + O(\omega_c^4\tau^4) \quad , \tag{4.5}$$

where

$$\sigma(\omega;0) = \frac{ne^2\tau}{m}[-i\omega\tau e^{1/\lambda} + \frac{1}{\lambda}(\omega\tau)^2 e^{2/\lambda} + O(\omega^3\tau^3)] \quad . \tag{4.6}$$

Eq.(4.5) shows that the magnetic field enhances the conductivity, although the conductivity vanishes at $\omega=0$. For larger values of ω and H, we must solve (4.3) and (4.4) numerically. The results are shown in Fig.3(a) where we have chosen λ to be 0.1 and g to be 2.

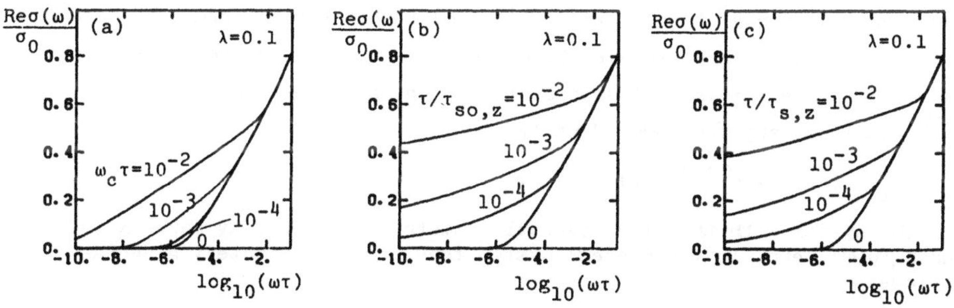

Fig.3 The conductivity $\mathrm{Re}\sigma(\omega)/\sigma_0$ vs. $\log\omega\tau$, where $\sigma_0=ne^2\tau/m$ and $\lambda=0.1$

(a) The system in a weak magnetic field, $\omega_c\tau=0$, 10^{-4}, 10^{-3} and 10^{-2}.

(b) The system with the spin-orbit scattering; $\tau/\tau_{so,z}=0$, 10^{-4}, 10^{-3} and 10^{-2}.

(c) The system with the paramagnetic impurities; $\tau/\tau_{s,x}=\tau/\tau_{s,z}=0$, 10^{-4}, 10^{-3} and 10^{-2}

4-2. Spin-orbit scattering and paramagnetic impurity scattering

When there are spin-orbit scattering or paramagnetic impurities, the scattering amplitudes depend on the spin states of particles. Hence $\phi_{pp'}(q,\omega)$, $\psi_{pp'}(q,\omega)$, $M_1(q,\omega)$ and $M_2(q,\omega)$ depend on spin states:

$$\phi_{pp'}^{\alpha\beta\gamma\delta}(q,\omega) \equiv -\frac{1}{2\pi i}\langle G_{\alpha\beta}^{R}(p_+,p'_+;E_F+\omega)\, G_{\gamma\delta}^{A}(p'_-,p_-;E_F)\rangle \quad , \tag{4.7}$$

$$\psi_{pp'}^{\alpha\beta\gamma\delta}(q,\omega) \equiv -\frac{1}{2\pi i}\langle G_{\alpha\beta}^{R}(p_+,p'_+;E_F+\omega) G_{\gamma\delta}^{A}(-p_-,-p'_-;E_F)\rangle \quad . \tag{4.8}$$

However the impurity averaged one-particle Green function $G^{R(A)}(p;\omega)$ does not depend on the spin state, since there is no magnetic field.

The derivation of the equations for $\phi^{\alpha\beta\gamma\delta}(q,\omega)$, and $\psi^{\alpha\beta\gamma\delta}(q,\omega)$ are essentially the same as before. The resulting equations are given as follows:

$$[\omega + K_1^{\alpha\alpha\delta\delta}(q,\omega)]\, \phi^{\alpha\beta\gamma\delta}(q,\omega) + K_1^{\alpha\bar{\alpha}\bar{\delta}\delta}(q,\omega)\phi^{\bar{\alpha}\beta\gamma\bar{\delta}}(q,\omega) - \phi_j^{\alpha\beta\gamma\delta}(q,\omega)$$
$$= N(E_F)\,\delta_{\alpha,\beta}\delta_{\gamma,\delta} \quad , \tag{4.9}$$

$$[\omega + M_1^{\alpha\alpha\delta\delta}(q,\omega)]\, \phi_j^{\alpha\beta\gamma\delta}(q,\omega) + M_1^{\alpha\bar{\alpha}\bar{\delta}\delta}(q,\omega)\, \phi_j^{\bar{\alpha}\beta\gamma\bar{\delta}}(q,\omega)$$
$$= (p_F^2q^2/2m^2)\phi^{\alpha\beta\gamma\delta}(q,\omega) \quad , \tag{4.10}$$

where

$$\begin{Bmatrix} K_1^{\alpha\beta\gamma\delta}(q,\omega) \\ M_1^{\alpha\beta\gamma\delta}(q,\omega)\end{Bmatrix} = \frac{i}{2\pi N(E_F)} \begin{Bmatrix} 1 \\ \dfrac{2(p\cdot q)(p'\cdot q)}{p_F^2 q^2}\end{Bmatrix} \Delta_+G_p\Delta_+G_{p'}$$
$$\times \{ -[\Sigma^R(p_+;E_F+\omega)-\Sigma^A(p_-;E_F)]\,\delta_{\alpha,\beta}\delta_{\gamma,\delta} + U_{pp'}^{\alpha\beta\gamma\delta}(q,\omega)\}, \tag{4.11}$$

The equations for $\psi^{\alpha\beta\gamma\delta}(q,\omega)$ and $\psi_j^{\alpha\beta\gamma\delta}(q,\omega)$ have the same form except that K_1, M_1, Δ_+G_p and U are replaced by K_2, M_2, Δ_-G_p and V, respectively. In these equations $\bar{\alpha}$ means the spin state orthogonal to α. The functions $K_1^{\alpha\beta\gamma\delta}(q,\omega)$ satisfy the Ward identity:

$$K_1^{\alpha\alpha\alpha\alpha}(q,\omega) + K_1^{\alpha\bar{\alpha}\bar{\alpha}\alpha}(q,\omega) = 0 \quad . \tag{4.12}$$

For the system with the time reversal symmetry this identity makes $K_1(q,\omega)$ and $K_2(q,\omega)$ zero. In the present system they are approximately given as follows:

$$K_1^{\alpha\alpha\alpha\alpha}(q,\omega) = -K_1^{\alpha\bar{\alpha}\bar{\alpha}\alpha}(q,\omega) = 2i\tau_{s,x}^{-1} \quad , \tag{4.13}$$

$$K_1^{\alpha\alpha\bar{\alpha}\bar{\alpha}}(q,\omega) = 2i(\tau_{s,z}^{-1} + \tau_{so,z}^{-1} + \tau_{s,x}^{-1}) \quad , \tag{4.14}$$

$$K_2^{\alpha\alpha\alpha\alpha}(q,\omega) = 2i(\tau_{so,z}^{-1} + \tau_{s,x}^{-1}) \quad , \tag{4.15}$$

$$K_2^{\alpha\bar{\alpha}\bar{\alpha}\alpha}(q,\omega) = 2i\,\tau_{s,x}^{-1} \quad , \tag{4.16}$$

$$K_2^{\alpha\alpha\bar{\alpha}\bar{\alpha}}(q,\omega) = 2i(\tau_{s,z}^{-1} + \tau_{s,x}^{-1}) \quad . \tag{4.17}$$

We approximate $U_{pp'}^{\alpha\beta\gamma\delta}(q,\omega)$ and $V_{pp'}^{\alpha\beta\gamma\delta}(q,\omega)$ by $\psi^{\alpha\beta\gamma\delta}(q,\omega)$ and $\phi^{\alpha\beta\gamma\delta}(q,\omega)$ as before. Then we get the following self-consistent equations:

$$\Phi_1(\omega) = 1+\lambda(1-2Z_{so}-2X_s)^2\Psi_1(\omega)\ell n\{1+[-i\omega\tau+2(Z_{so}+X_s)]^{-1}\Psi_1(\omega)^{-1}\} \quad , \qquad (4.18)$$

$$\Phi_2(\omega)\pm\Phi_3(\omega) = 1+\lambda[1-2Z_s-(2\mp2)X_s]^2[\Psi_2(\omega)\pm\Psi_3(\omega)] \times \ell n\{1+[-i\omega\tau+2Z_s+(2\mp2)X_s]^{-1}[\Psi_2(\omega)\pm\Psi_3(\omega)]^{-1}\} \quad , \qquad (4.19)$$

$$\Psi_1(\omega)\pm\Psi_3(\omega) = 1+\lambda[1-(2\mp2)X_s]^2[\Phi_1(\omega)\pm\Phi_3(\omega)] \times \ell n\{1+[-i\omega\tau+(2\mp2)X_s]^{-1}[\Phi_1(\omega)\pm\Phi_3(\omega)]^{-1}\} \quad , \qquad (4.20)$$

$$\Psi_2(\omega) = 1+\lambda(1-2Z_s-2Z_{so}-2X_s)^2\Phi_2(\omega)\ell n\{1+[-i\omega\tau+2Z_s+2Z_{so}+2X_s]^{-1}\Phi_2(\omega)^{-1}\} \quad , \qquad (4.21)$$

where $\Phi_1(\omega)=-i\tau M_1^{\alpha\alpha\alpha\alpha}(0,\omega)$, $\Phi_2(\omega)=-i\tau M_1^{\alpha\alpha\bar{\alpha}\bar{\alpha}}(0,\omega)$, $\Phi_3(\omega)=-i\tau M_1^{\alpha\bar{\alpha}\bar{\alpha}\alpha}(0,\omega)$, $\Psi_1(\omega)=-i\tau M_2^{\alpha\alpha\alpha\alpha}(0,\omega)$, $\Psi_2(\omega)=-i\tau M_2^{\alpha\alpha\bar{\alpha}\bar{\alpha}}(0,\omega)$, $\Psi_3(\omega)=-i\tau M_2^{\alpha\bar{\alpha}\bar{\alpha}\alpha}(0,\omega)$, $X_s=\tau/\tau_{s,x}$, $Z_s=\tau/\tau_{s,z}$, $X_{so}=\tau/\tau_{so,z}$, and $\tau^{-1}=\tau_0^{-1}+\tau_{so,z}^{-1}+2\tau_{s,x}^{-1}+\tau_{s,z}^{-1}$. We notice that when X_s or Z_{so} is finite, there is no $\ell n(\omega\tau)$ term in $\Phi_1(\omega)$ and $\Phi_3(\omega)$. Hence the conductivity, which is given by

$$\sigma(\omega) = \frac{ne^2\tau}{m}[-i\omega\tau + \Phi_1(\omega) + \Phi_2(\omega)]^{-1} \quad , \qquad (4.22)$$

remains finite. On the other hand when only τ_0 and Z_s is finite, the diffusion pole is not suppressed and $\sigma(0)$ is zero. This case is the same as that of normal impurities only, except that $\tau^{-1}=\tau_0^{-1}+\tau_{s,z}^{-1}$ should be used instead of τ_0^{-1}.

The conductivity at finite frequency must be calculated numerically. The results are shown in Fig.3(b) and 3(c) for the case of spin-orbit scattering and paramagnetic impurity scattering respectively. In these figures we have chosen λ to be 0.1.

5. Discussion

In this paper we have calculated the dynamical conductivity $\sigma(\omega)$ diagrammatically. We have considered two-dimensional electrons in weak perturbations which destroy the time reversal symmetry besides the normal impurity potential. Our method is a refined version of that by VOLLHARDT and WÖLFLE [4]. The conductivity is expressed by the current relaxation kernel $M_1(q,\omega)$. To calculate $M_1(q,\omega)$ we have approximated the irreducible vertices $U_{pp'}(q,\omega)$ and $V_{pp'}(q,\omega)$ by the diagrams shown in Fig.2, and obtained coupled self-consistent equations for $M_1(q,\omega)$ and for another relaxation kernel $M_2(q,\omega)$. This approximation is good when $\lambda<<1$ and the perturbation that destroys the time reversal symmetry is weak. When $\lambda\gtrsim1$, more complicated diagrams will contribute to $U_{pp'}(q,\omega)$ and $V_{pp'}(q,\omega)$ in the equal order of magnitude as those in Fig.2, and the approximation becomes inadequate. The approximation to use $\psi(q,\omega)$ for $U_{pp'}(q,\omega)$ is justified when $\psi(q,\omega)$ has a structure like a diffusion pole. In the presence of the time reversal symmetry $\psi(q,\omega)$ does have a diffusion pole. However the perturbations which destroy the symmetry suppress the pole. Hence it is necessary to have $\omega_c\tau \ll 1$, $\tau_{s,i} \gg \tau$ and $\tau_{so,z} \gg \tau$. When these conditions are not satisfied, the diagrams which contain $\psi(q,\omega)$ become relatively unimportant for $U_{pp'}(q,\omega)$ in comparison with the other diagrams.

We get the conductivity by solving these self-consistent equations. The static conductivity remains finite when there is spin-orbit scattering or spin-flip scattering by the paramagnetic impurities. On the other hand the static conductivity vanishes in the absence of these scatterings, even if there is a weak magnetic field.

VOLLHARDT and WÖLFLE [5] have shown that their theory is consistent with the single parameter scaling theory by ABRAHAMS et al. [1], and they actually obtained the β-function of [1]. This is because they considered a system with normal impurities only. For such a system there is the time reversal symmetry and the conductivity is determined by only one relaxation kernel $M_1(q,\omega)$ which coincides with $M_2(q,\omega)$ due to the symmetry. However for the systems that we considered in this paper more than one relaxation kernels are necessary, and we cannot get a β-function which obeys the single parameter scaling equation. We will need more than one scaling parameters to describe the scaling equation, and the single parameter scaling theory of [1] does not give the correct behavior of the conductivity in the absence of the time reversal symmetry.

We have treated the impurity scattering by the Born approximation. Since we assume $\lambda \ll 1$, this approximation is generally good. However the scattering by the paramagnetic impurities may be exceptional. In pure three-dimensional system the effect of a paramagnetic impurity on the conduction electron has been studied as the Kondo problem. It is known that in such a system the Born approximation is not good. It is possible to improve our theory by considering $\tau_{s,i}$'s as some effective relaxation times, although we do not know what value should be used for $\tau_{s,i}$'s. What our theory tells us is that when the spin-flip scattering is possible, the static conductivity becomes finite. OHKAWA [11] considerded the Kondo problem in dirty two-dimensional system. According to his theory there is no spin-flip scattering in the ground state, and the static conductivity becomes zero.

We have considered a strictly two-dimensional model. Hence there is no spin-flip scattering due to the spin-orbit scattering. It is possible to consider a model where $\tau_{so,x}^{-1}=\tau_{so,y}^{-1}\neq 0$. If we apply our method to this model, the static conductivity diverges. This means that the model is unphysical. In physical system the finite value of $\tau_{so,x}^{-1}$ and $\tau_{so,y}^{-1}$ requires the finite thickness of the system, and there must be many subbands due to the motion along the z-axis. To be applicable to such many-subband system our theory must be extended and the relaxation kernels which depend on the subband indices should be introduced.

Acknowledgement

This work was done in colaboration with Professor H. Fukuyama and Dr. Y. Ono. The author thanks them.

References

1. E. Abrahams, P.W. Anderson, D.C. Licciardello and T.V. Ramakrishnan, Phys. Rev. Lett. 42, 673 (1979).
2. E. Abrahams and T.V. Ramakrishnan, J. Non-Crys. Sol. 35, 15 (1980).
3. L.P. Gor'kov, A.I. Larkin and D.E. Khmelnitzkii, JETP Lett 30, 248 (1979).
4. D. Vollhardt and P. Wölfle, Phys. Rev. B22, 4666 (1980).
5. D. Vollhardt and P. Wölfle, preprint.
6. P.A. Lee, J. Non-Crys. Sol. 35, 21 (1980).
7. S. Hikami, A.I. Larkin and Y. Nagaoka, Prog. Theor. Phys. 63, 707 (1980).
8. Y. Ono, D. Yoshioka and H. Fukuyama, J. Phys. Soc. Japan 50, 2143 (1981).
9. D. Yoshioka, Y. Ono and H. Fukuyama, J. Phys. Soc. Japan 50, 3419 (1981).
10. D. Yoshioka, J. Phys. Soc. Japan 51 No.3 (1982).
11. F.J. Ohkawa, these conference proceedings.

Localization in One-Dimensional and Quasi-One-Dimensional Systems

A. MacKinnon

Universität Heidelberg, Institut für Theoretische Physik
Philosophenweg 19, D-6900 Heidelberg, Fed. Rep. of Germany

The localisation length for disordered systems is calculated by a new recursive method using long strips or bars to describe 2-and 3-dimensional systems, respectively. By studying their behaviour as a function of cross-section, scaling relationships for localisation length and conductance are derived. The assumptions about the properties of the ß-function used in recent analytical work are confirmed.

1. Review of Numerical Methods

In the last ten years a number of numerical techniques have been developed to study the problem of Anderson localisation in disordered systems. Since traditional diagonalisation methods are restricted to systems of the order of 100 sites (see Table 1) it was found necessary to look for calculational methods which could deal with larger systems. These methods either made use of the sparse nature of the Hamiltonian matrix [1,2] in the Anderson model or directly calculated some average quantity without explicitly using the eigenfunctions [2,3,4]. It was thus possible to calculate quantities such as the participation ratio and the conductivity of systems of up to 10^4 sites; the restriction again being due to computer storage.

The increasing use of renormalisation group ideas [5,6] in analytical work led to a number of attempts to look for scaling relationships numerically [7-10].

In most of the above work, one-dimensional systems were not considered seriously since it is well known that all states are localised in this case [11]. However, the observation by CZYCHOLL and KRAMER [12] that this "trivial" result is difficult to reproduce numerically, led to the development of a number of new methods which make use of the special topological properties of 1D systems [13-16]. These methods are generally not limited by computer storage but rather by time (Table 1).

The results of these calculations and of analytical work on 1D systems [16-18] led to the realisation that the statistical fluctuations between different realisations of a system are such that it is necessary to average over a very large number of systems to achieve acceptable accuracy.

Thus many previous results are not very conclusive, either because the systems are too small or because the error bars are large. In the case of calculations of the conductivity using the KUBO-GREENWOOD formula statistical accuracy could only be achieved by introducing an energy level broadening, which tends to smooth away the critical behaviour [13, 17].

Table 1 Computer requirements of computational methods

Method	Storage	Time
Diagonalisation	N^2	N^3
Sparse Matrix Methods [1,2]	N	N^2
Averaging Methods [2-4]		
Renormalisation and Decimation [7-10]	<N	N^2
One Dimensional Methods [13-18]	1	N

2. Method of Calculation

The above discussion suggests that one should attempt to exploit the advantages of 1D methods in the study of 2D and 3D systems. The most obvious way of implementing this concept is to consider 2 and 3 dimensional systems where one of the dimensions is much larger than the others, i.e. an NxM or NxMxM system where N >> M. We should expect to be able to glean information about 2D and 3D systems by studying the dependence of the properties of such long strips or bars on the thickness M.

In this paper I shall concentrate on the Green's function or resolvent matrix elements between sites at opposite ends of the system, i.e. $\langle i,1|(Z-H)^{-1}|j,N\rangle$ where the indices "i" and "j" run over the M^{d-1} sites in the cross section and "1" and "N" are the ends. In what follows I shall drop "i" and "j" and write $\langle 1|G|N\rangle$. This matrix can be calculated in several ways, for example:

$$\begin{aligned}\langle 1|G|N+1\rangle &= \langle 1|G|N\rangle\langle N+1|G|N+1\rangle \\ \langle N+1|G|N+1\rangle &= \left(E - H_{N+1} - \langle N|G|N\rangle\right)^{-1}\end{aligned} \qquad (1)$$

where H_{N+1} is the Hamiltonian of the (N + 1)st isolated chain or sheet, and the units have been chosen so that the off diagonal elements of the Anderson Hamiltonian are equal to unity. In (1) and in what follows I have used the Anderson [24] Hamiltonian with diagonal rectangular disorder, $-\frac{W}{2} < \varepsilon < \frac{W}{2}$, and the units have been chosen so that the off diagonal elements V between nearest neighbours are equal to unity. In all cases the energy E is taken as zero and periodic boundary conditions are used in the second and third dimensions. An alternative procedure is

$$\begin{aligned}A_{N+1} &= (E - H_{N+1})A_N - A_{N-1} \\ \langle 1|G|N\rangle &= A_N^{-1}\ .\end{aligned} \qquad (2)$$

Both these procedures have disadvantages. Procedure (1) requires a matrix inversion at each step, which is computationally costly. On the other hand procedure (2) is numerically unstable. The eigenvalues of A_N rise exponentially with N at different rates. When the ratio of the smallest to the largest eigenvalue becomes comparable with the accuracy of our machine the smallest is lost. This smallest eigenvalue however provides the dominant contribution to $\langle 1|G|N\rangle$.

By combining the above procedures we have developed a method which requires a matrix inversion only after a number of steps slightly less than that at which the smallest eigenvalue of A_N would be lost [19].

An alternative more costly method [20] involves frequent diagonalisation of the transfer matrix

$$\prod_N \begin{bmatrix} (E - H_N) & -I \\ I & 0 \end{bmatrix} .$$

Assuming that <1|G|N> has the form

$$\mathrm{Tr}|\langle 1|G|N\rangle|^2 \sim \exp -2N/\lambda_M(W) \tag{3}$$

for large N, we are able to calculate the localisation length $\lambda_M(W)$ from

$$\frac{1}{\lambda_M(W)} = \lim_{N\to\infty} \frac{1}{2N} \sum_{k=2}^{N} \ln \frac{\mathrm{Tr}\langle 1|G|N-1\rangle}{\mathrm{Tr}\langle 1|G|N\rangle} . \tag{4}$$

This function has the form of an average. The corresponding variance can be used to determine the accuracy with which $\lambda_M(W)$ has been calculated. We can thus continue making the system longer until a predetermined condition has been fulfilled, in our case $\Delta\lambda_M(W) < 0.01\ \lambda_M(W)$.

Considering the system as a 1D chain of "block spins" [21] of size M^d we find the renormalised localisation length $\lambda'_M = \lambda_M/M$. In Fig.1 this quantity is plotted against M for various values of disorder, W . In 2D (Fig.1a) λ'_M clearly gets smaller as M gets larger for all W, whereas in 3D (Fig.1b)

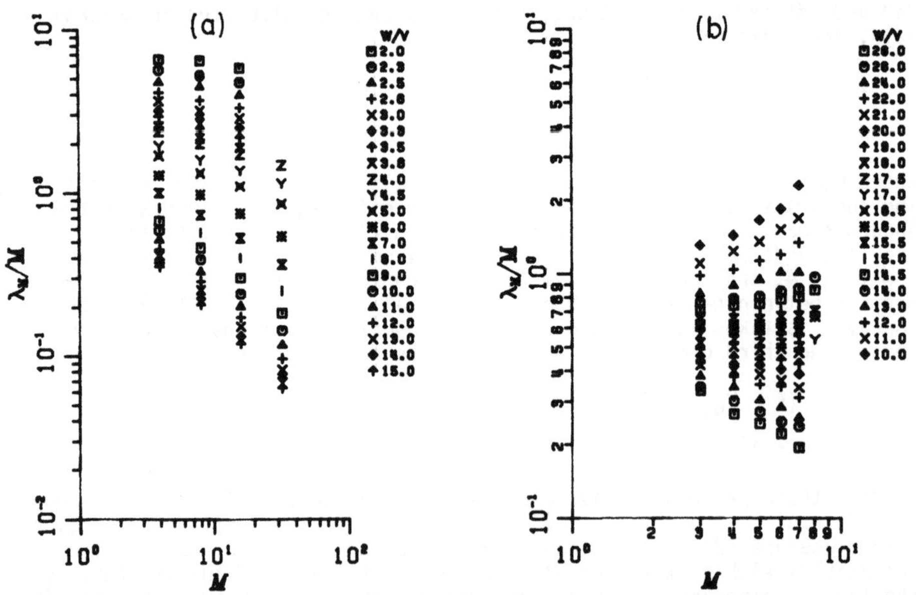

Fig.1 Renormalised localisation length λ_M/M vs. system width M for various values of disorder, W/V. (a) 2D (b) 3D

there is a transition from such behaviour to a regime where λ'_M rises with M. Since the statistical error in these results is comparable with the size of the symbols it is difficult to pinpoint the critical disorder W_c much better than 14 V < W_c < 18 V. Thus we have no evidence of critical behaviour in two dimensions, whereas in three dimensions we observe an Anderson transition.

3. Scaling Behaviour

3.1 Localisation Length

In order to understand these results better we look for scaling behaviour of the form

$$\lambda'_M = \lambda_M/M = f_d(\lambda_\infty/M) \tag{5}$$

where λ_∞ depends only on the disorder. This is in practice equivalent to the assumption that the scaling behaviour of λ'_M depends only on λ'_M itself, i.e.

$$d \ln \lambda'_M / d \ln M = F_d(\lambda'_M) \; . \tag{6}$$

We determine the λ_∞'s in (5) by minimizing the variance of the various values of $\ln(\lambda_\infty/M)$ which correspond to each λ'_M. In this way we can determine the λ_∞'s and thus $f_d(x)$, except for a coefficient. However, since we expect that in the localised regime $\lim_{x\to 0} f_d(x) = x$, i.e. $\lim_{M\to\infty} \lambda_M = \lambda_\infty$, we can

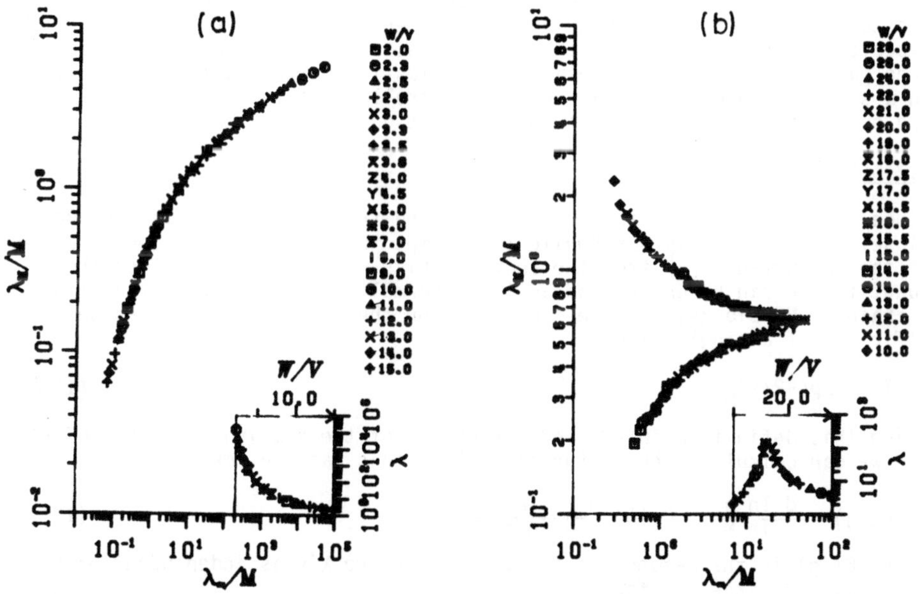

Fig.2 Renormalised localisation length λ_M/M vs. λ_∞/M for various values of disorder W/V, i.e. $\lambda_M/M = f_d(\lambda_\infty/M)$. (a) 2D (b) 3D

identify λ_∞ with the localisation length of the 2D or 3D system and fix the unknown coefficient by assuming that the smallest λ'_M has already converged to λ_∞/M.

The functions $f_d(x)$ and the λ_∞'s are shown in Fig.2. The scaling law (5) is clearly very well obeyed in both 2D and 3D. In 2D there is again no sign of critical behaviour, whereas in 3D there is a fixed point at $\lambda'_c \sim 0.6$.

What is the significance of the characteristic length λ_∞ for the upper branch in 3D (Fig.2b)? It clearly cannot be a localisation length. In order to understand this we first note that, for large M, we have

$$\lambda_M/M \sim M/\lambda_\infty \ . \tag{7}$$

Since for very long systems $<1|G|N>$ has one dominant eigenvalue we can write an effective transmission coefficient

$$T = \exp(-2N/\lambda_M) \ . \tag{8}$$

Assuming that the LANDAUER [22] formula for the conductance of a 1D system is also valid in this limit we can write

$$g = \frac{T}{1 - T} \ . \tag{9}$$

Combining (7), (8) and (9) and taking the limit of large g we have

$$g \sim M^2/N \lambda_\infty \tag{10}$$

Thus λ_∞ is nothing other than the resistivity. The result is significant for a number of reasons. The authors of [6], the "gang of four", were unable to show that the mean free path is not a relevant length. We now have the result (10) that the relevant length in the conducting regime is related to the conductivity "σ" by $\sigma \sim \lambda^{-1}$ rather than by $\sigma \sim \lambda$ as would be expected for the mean free path. This suggests that the concept of "mean free path", which actually relates to inelastic scattering is not relevant in a situation where the transport is controlled solely by elastic scattering processes, but rather that the relevant quantity is the resistivity.

3.2 Conductance

Since, as we have seen, the localisation and transport properties are very closely related, it is worthwhile to attempt to derive the scaling behaviour of the conductance [6] from the above results. As mentioned earlier the quantity λ'_M can be understood as the renormalised localisation length of a chain of M^d blocks. Thus we can write for the effective transmission coefficient of a single block

$$T = \exp(-2/\lambda'_M) \ . \tag{11}$$

Using this definition in (9) and making use of the scaling behaviour of λ'_M (5) we can calculate the "ß-function" [6]. This has the form

$$\beta_d(g) = \frac{d \ln g}{d \ln M} = \beta_1(g) \frac{d \ln f_d(x)}{d \ln x} \ , \tag{12}$$

where $\beta_1(g)$ is the ß-function for the case where λ'_M is independent of M, i.e. 1D, and takes the form

$$\beta_1(g) = (1 + g) \ln (g/(1 + g)) \ . \tag{13}$$

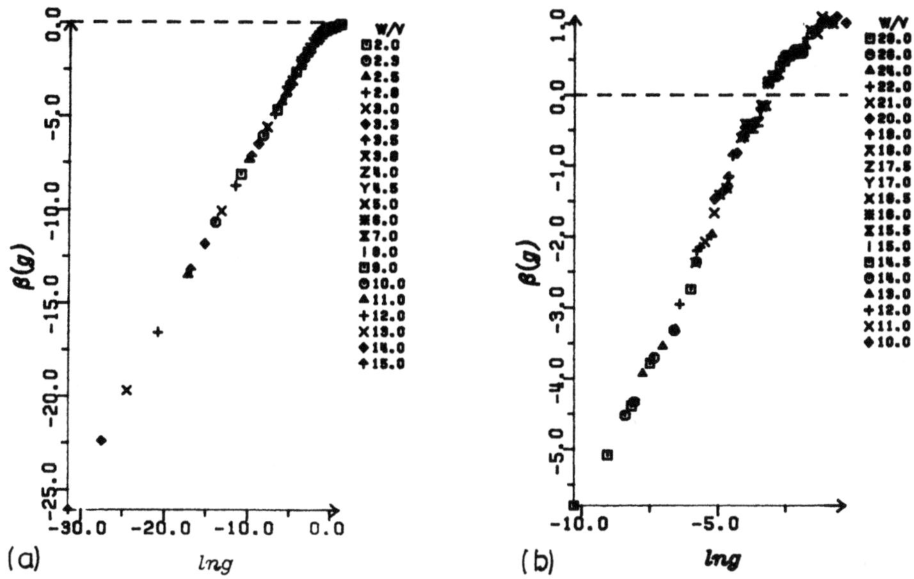

Fig.3 β-function, d ln g/d ln M vs. conductance (g). (a) 2D (b) 3D

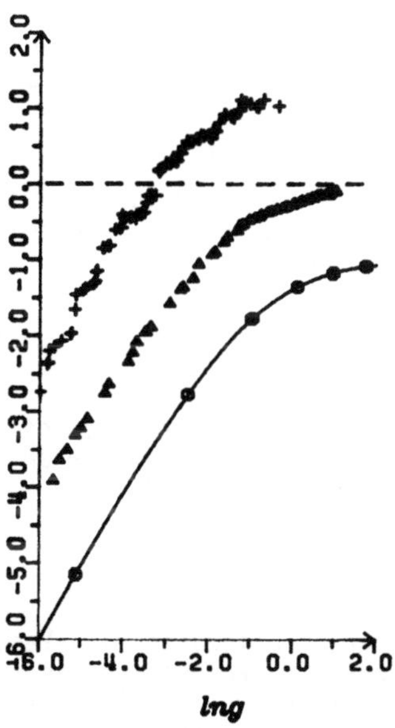

Fig.4 β-function near β = 0.
○ 1D; ▲ 2D; + 3D

These scaling functions are shown in Fig.3, and the critical region, around β = 0 is shown on a larger scale in Fig.4. These β-functions agree very well with those derived analytically by VOLLHARDT and WÖLFLE [23] except that the 2D and 3D functions are shifted slightly along the ln g axis. This agreement is very encouraging since our calculation of $\beta_d(g)$ involves a not insignificant approximation. The replacement of $<1|G|N>$ or the transmission matrix by its largest eigenvalue is only valid in the limit $N \to \infty$. For $N = M$, i.e. square or cubic blocks, the contributions from the other, smaller, eigenvalues can no longer be ignored. The approximation we have made is however to ignore these smaller eigenvalues. This is equivalent to using the known scaling behaviour of truly 1D systems (13) to scale the long dimension "N" of the system back to $N = M$. Another way of looking at this single channel approximation is that it is equivalent to replacing the "spins" on an M^d block with a single "block spin" as in the conventional scaling theory of phase transitions [21]. This last argument helps to explain why the functional form of our scaling functions agrees with [23], whereas there are quantitative differences. It is, however, also possible that the differences are related to different definitions of the conductance.

3.3 Critical Indices

For $d = 3$ the critical indices ν and s, for the localisation length and the conductivity respectively, can be calculated either directly from $\lambda_\infty(W)$ or indirectly from $d\beta(g)/d \ln g$ at $\beta = 0$ [6]. In both cases they are found to be equal and $s = \nu = 1.2 \pm 0.3$, the error bar resulting mainly from the uncertainty in the position of W_c or from the related difficulty in determining the behaviour of $f_d(x)$ near the fixed point (Fig.2b). Nevertheless, we can determine W_c from the scaling behaviour rather more accurately than from the raw data. Thus $W_c = 16 \pm 1$ V.

We consider these results as in good agreement with [23] who derived analytically $s = \nu = 1$.

For $d = 2$ we find it very difficult to assign any power law behaviour for $\lambda_\infty(W)$. A fit to an essential singularity of the kind proposed in [23] turned out to be successful implying that $\lambda_\infty(W) \sim \exp(\text{const}/W^2)/W^2$. However, the data are also consistent with $\lambda_\infty(W) \sim \exp(\text{const}/W)$.

We were unable to fit the sort of behaviour found by PICHARD and SARMA [20] although our raw data look very similar. We believe this is because we have calculated a larger range of values of both W and M.

4. Conclusion

We have developed a method of calculating the localisation properties of long quasi-one-dimensional systems described by the Anderson Hamiltonian, with an accuracy which is much better than has been possible hitherto.

We have shown that the dependence of the localisation length on the thickness of the strip or bar obeys a scaling law.

For $d < 2$ we found that all states are localised. The localisation length is of the order of 10^3 at $W = 4$ V and clearly reaches macroscopic sizes for $W > V$. We are thus in a position to understand why previous calculations have tended to find an Anderson transition at $W \sim 6$ V: the localisation length becomes comparable with system sizes used.

For $d = 3$ on the other hand we observe an Anderson transition at $W_c \sim 16$ V, where it is traditionally found.

In all cases considered we have observed a scaling behaviour that depends on only one parameter, as predicted by [6]. This parameter can be considered to be the dimensionless conductance "g" or the renormalised localisation length λ'_M, according to taste.

We have seen no evidence of power law localisation, neither in 2D nor in 3D [20]. Our scaling functions are smooth everywhere; except possibly at $\beta_3 = 0$, where the fixed point in $f_3(x)$ makes this difficult to prove. On the other hand the fact that our critical indices s and ν are equal, as far as we can see, is consistent with a smooth β-function.

We have derived the scaling behaviour without any a priori assumptions about the form of $f_d(x)$, $\beta_d(g)$ or $\lambda_\infty(W)$ solely from the data. Our only approximation is the consideration of a single channel in the calculation of $\beta_d(g)$.

Our results agree fully with the predictions of the "gang of four" [6] and can be considered an a posteriori confirmation of their assumptions of continuity and monotonicity.

References

1. S. Yoshino, M. Okazaki: J. Phys. Soc. Jap. 43, 415 (1977)
2. J. Stein, U. Krey: Z. Physik B34, 287 (1979); ibid B37, 13 (1980)
3. D. Weaire, A. R. Williams: J. Phys. C10, 1239 (1977);
   D. Weaire, V. Srivastava: Solid St. Commun. 23, 863 (1977)
4. B. Kramer, A. MacKinnon, D. Weaire: Phys. Rev. B23, 6357 (1981)
5. F. Wegner: Z. Physik B25, 537 (1976);
   R. Oppermann, F. J. Wegner: Z. Physik B34, 327 (1979)
6. E. Abrahams, P. W. Anderson, D. C. Licciardello, T. V. Ramakrishnan: Phys. Rev. Lett. 42, 673 (1979)
7. P. A. Lee: Phys. Rev. Lett. 42, 1492 (1979)
8. S. Sarker, E. Domany: Phys. Rev. B23, 6018 (1981); J. Phys. C13, L273 (1980)
9. H. Aoki: J. Phys. C13, 3369 (1980)
10. C. Lambert, D. Weaire: to be published
11. N. F. Mott, W. D. Twose: Adv. Phys. 20, 107 (1961)
12. G. Czycholl, B. Kramer: Z. Physik B39, 193 (1980)
13. A. MacKinnon: J. Phys. C13, L1031 (1980)
14. D. J. Thouless, S. Kirkpatrick: J. Phys. C14, 235 (1981)
15. E. N. Economou, C. M. Soukoulis: Phys. Rev. Lett. 46, 618 (1981)
16. J. Sak, B. Kramer: Phys. Rev. B24, 1761 (1981)
17. G. Czycholl, B. Kramer, A. MacKinnon: Z. Physik B43, 5 (1981)
18. P. W. Anderson, D. J. Thouless, E. Abrahams, D. S. Fisher: Phys. Rev. B22, 3519 (1980)
19. A. MacKinnon, B. Kramer: Phys. Rev. Lett. 47, 1546 (1981)
20. J. L. Pichard, G. Sarma: J. Phys. C14, L127 (1981)
21. L. P. Kadanoff: Physics 2, 263 (1966)
22. R. Landauer: Phil. Mag. 21, 863 (1970)
23. D. Vollhardt, P. Wölfle: Phys. Rev. B22, 4666 (1980) and to be published (see this volume)
24. P. W. Anderson: Phys. Rev. 109, 1492 (1958)

Numerical Studies of Localization in Two Dimensions

P.A. Lee

Bell Laboratories
Murray Hill, NJ 07974, USA

ABSTRACT

We review the recent work of Lee and Fisher in which the conductance for a two dimensional tight binding model with on site disorder is calculated numerically using the Kubo formula. For weak disorder logarithmic localization is observed, in agreement with the scaling theory. The magnetoresistance is found to be negative in both the logarithmic and exponential localization regimes. Results for a model with random hopping matrix elements are also presented.

In the last two years, significant advances have been made in understanding the scaling behavior of Anderson localization [1,2]. Much of the work has been based on the idea [3] that the behavior as a function of length scale L is determined by a single scaling variable, the dimensionless conductance at scale L, $g(L) = \Gamma(L)/(e^2/\hbar)$ where $\Gamma(L)$ is the conductance of a system with linear dimension L. The renormalization group differential recursion relation which describes the changes in g as a function of length scale can be written.

$$\frac{d\ell ng}{d\ell nL} = \beta(g) \ . \tag{1}$$

The β function has been computed in two dimensions by perturbation theory in powers of g^{-1} for weak disorder. For spinless electrons it is found that $\beta(g) = -(2\pi^2 g)^{-1} + 0(g^{-4})$ where the absence of the zeroth order term is unique to 2D and suggests that two is the lower critical dimensionality for the Anderson transition. In 2D, localization is predicted for any amount of disorder, since g always decreases as a function of L corresponding to insulating behavior. For weak disorder, however, the effects of localization will only decrease g(L) locarithmically.

In this paper we review a recent work [4,5] which computes the conductance of $L \times L$ samples using a method based on the Kubo formula. A somewhat different approach based on the scaling behavior of long strips has been successfully implemented by PICHARD and SARMA [6] and by MACKINNON and KRAMER [7]. This latter approach is reviewed elsewhere in these proceedings.

We consider a tight binding Anderson model on a 2D square lattice,

$$H = \sum_{j,k} V_x^{jk} a_{j+1,k}^{+} a_{j,k} + V_y^{jk} a_{j,k+1}^{+} a_{j,k} + \text{c.c.}$$

$$+ \sum_{j,k} E_{jk} a_{j,k}^{+} a_{jk} \qquad (2)$$

where j,k label the sites in the x and y directions and E_{jk} is a random site energy distributed uniformly between $\pm W/2$. We initially take $V_x = V_y = 1$. The parameter W thus provides a measure of the disorder. The frequency dependent conductance of an L×L sample is given by the Kubo formula

$$\Gamma(\omega) = \frac{\pi}{\omega L^2} \int dE' \sum_{\alpha,\beta} \left| \sum_j < \alpha | J(j) | \beta > \right|^2 \delta(E'+\omega-E_\beta)\delta(E'-E_\alpha) \qquad (3)$$

where the current operator $J(j) = ie\sum_k V_x^{jk} (a_{j,k}^{+} a_{j-1,k} - \text{c.c.})$ and $|\alpha>$, $|\beta>$ are the eigenstates of the system with $E_\alpha < E < E_\beta$ where E is the Fermi energy. For an isolated sample the eigenvalues are discrete and $\Gamma(\omega)$ consists of a series of δ functions so that some procedure for averaging over ω must be introduced [8]. Furthermore the numerical computation of the current matrix elements is time consuming. We thus take a somewhat different approach by extending the disordered sample to infinity in the $\pm x$ directions by affixing an ordered tight binding lattice on each side. We take periodic boundary conditions in the y direction for all x, and apply an electric field in the x direction *only* in the disordered region. The energy spectrum will now be continuous and the $\omega \to 0$ limit of the conductance can be taken. Current conservation immediately implies that the matrix element $<\alpha|J(j)|\beta>$ is independent of j. The sum over j in (3) is hence trivially done.

We introduce the Green's functions $G^{(\pm)} = (E-H\pm i\eta)^{-1}$ and $2iG'' = G^{(-)} - G^{(+)}$. Defining $\Gamma = \Gamma(\omega{=}0)$, (3) simplifies to

$$\Gamma = (e^2/\pi)\mathrm{Tr}[G''(j_0,j_0')G''(j_0'-1,j_0-1) + G''(j_0-1,j_0')G''(j_0',j_0)$$

$$- G''(j_0,j_0'-1)G''(j_0',j_0-1) - G''(j_0-1,j_0')G''(j_0'-1,j_0)] \qquad (4)$$

where the trace is over the site index k and j_0,j_0' are arbitrary. By choosing j_0 and j_0' to be on opposite sides and far away from the disordered region it can be shown [5] that (4) is equivalent to

$$\Gamma = e^2/(2\pi\hbar)\mathrm{Tr}(t^{+}t) \qquad (5)$$

where t is the transmission matrix relating incoming waves on one side of the disordered region to "transmitted" waves on the other side. Eq. (5) is a generalization to multi-channel of the single channel result of ECONOMOU and SOUKOULIS [9]. One can compute t using a transfer matrix technique which is in principle more efficient than calculating the Greens's functions in (4), (the cost is αL^3 vs L^4) but the matrix inversion necessary to obtain t rapidly becomes singular, limiting this method to relatively small L, but providing a useful check to the calculations described below.

We should point out a shortcoming of our definition of the conductance. According to [5], in the absence of disorder, $\Gamma = (e^2/2\pi\hbar)N$ where N is the number of channels. It is only in the $N \to \infty$ limit that the expected result of $\Gamma^{-1} = 0$ is recovered. This difficulty can be traced to our definition which restricts the driving electric field to a region of size L. Thus conductivity at finite q is mixed in. An ideal system driven in this way will indeed be dissipative. This difficulty becomes more apparent if we consider a charged system (instead of a neutral system driven by chemical potential difference). By applying a field E only in the disordered region a current is generated in the ideal "wire" that is spatially nonuniform, i.e.

$$J(x,t) = \frac{ELe^2}{2\pi\hbar}\sum_{ab}|t_{ab}|^2 \exp(i\omega|x|/v_a - i\omega t)$$

where v_a is the velocity in channel a. This gives rise to density fluctuations that must be screened out in a charged system. A more satisfactory definition of conductance is to require that the current be uniform, and to solve for the electronic field. This was done recently [10,11] in the single channel case, where it was shown that a spatially uniform out of phase field in the wire produces a uniform total current, and the formula $\Gamma = (e^2/2\pi\hbar)\ |t|^2/(1-|t|^2)$ first written down by Landauer [12] was recovered. A similar strategy would work in the multi-channel case as well [11], but a simple expression analogous to the Landauer formula does not exist. It is expected that in the limit of large N, the various definitions of conductance should converge and (5) is adequate. In defining the resistance for a finite system, there is always ambiguity in the treatment of boundary scattering which can be of order L^{-1}. This ambiguity is comparable to the difference which is of order N^{-1} in the ideal limit as given by (5). It will be interesting in the future to implement the Abrahams-Langreth [11] definition of conductance and see if that will lead to an improvement in the convergence with sample size.

The Green's functions required in (4) are computed by a recursive method which builds up the sample in the x direction one column at a time, starting from the left with a system in which all sites $j > j_0$ are deleted. The Green's function $G^L(j_0) \equiv G^L_{kk'}(j_0,j_0)$ for the last column obeys the matrix recursion relation

$$G^L(j_0) = [G^0(j_0)^{-1} - V_x^2 G^L(j_0-1)]^{-1} \tag{6}$$

where $G^0(j_0)$ is the Green's function for the isolated j_0th column. Similarly we introduce $G^R(j_0)$ as the first column Green's function for a system starting at j_0. It is easy to show that

$$G(j,j) = [G^0(j)^{-1} - V_x G^R(j+1) - V_x^2 G^L(j-1)]^{-1} \tag{7}$$

and

$$G(j,j+1) = G(j,j)V_x G^R(j+1) \ . \tag{8}$$

We have assumed V_x to be a constant in the above discussion, but this condition can easily be relaxed.

Initially G^L is computed for a semi-infinite system of a perfect lattice and then by repeated use of (6), G^L is computed for the left side plus the disordered region. Finally another semi-infinite perfect system is attached to the right using (7,8). The conductance is then calculated from (4) with $j_0 = j_0'$ just outside the disordered region.

The logarithm of the dimensionless conductance is computed for M different samples with a fixed W and E. The mean $\overline{\ln g}$ and the rms deviation $\Delta \ln g$ are then calculated with typically M=200 for L=16 and 32 and M=100 for L=64. The results for W = 4, E = 0 (band center) are shown in the insert in Fig. 1.

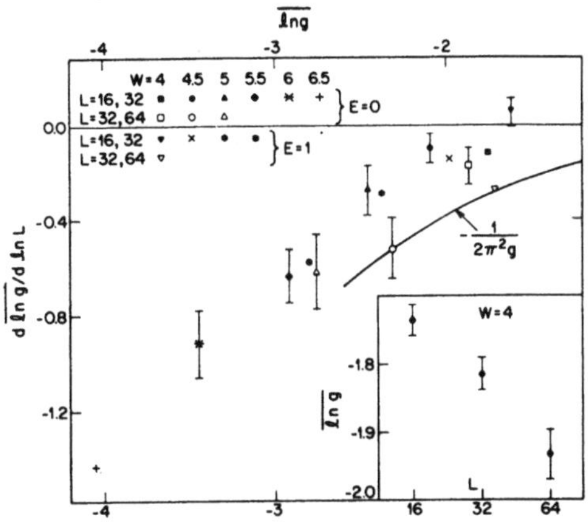

Fig. 1. Numerically calculated scaling function for the Anderson model showing the dependence of the logarithm of the conductance g on length scale L

The error bars are statistical, given by $\pm \Delta \ln g/\sqrt{M}$. We see that $\ln g$ decreases by roughly 15% per doubling of L in at least qualitative agreement with the logarithmic localization predicted by the scaling theory. For the same value of W, the previous calculation by LEE [13] showed metallic behavior, i.e. $\overline{\ln g}$ independent of L. That calculation was approximate in that the basis set was truncated with each doubling of L, and can be viewed as an approximation valid for strong disorder. Apparently systematic errors were sufficiently large at W = 4 to produce the wrong answer. It is worth noting that STEIN and KREY [8] computed the conductance using the Kubo formula for isolated samples and obtained $\ln g = -1.65$ for W = 4 independent of L. Due to various approximations their calculations are effectively cut-off at relatively small L, so that the agreement with the present result is reasonable. We further note that (5) most likely underestimates the conductance and requires corrections of order ℓ/L (ℓ is the mean free path) to account for boundary resistance and the effects of current in the perfect regions. Since these corrections would probably lead to smaller underestimates of the conductance for larger systems, we believe that our observed decrease in $\overline{\ln g}$ with increasing L is a slight underestimate.

By calculating the differences $\overline{\ln g(2L)} - \overline{\ln g(L)}$ for various values of W we construct the numerical scaling function shown in Fig. 1. The open symbols are obtained from L = 64 and L = 32 samples and are quite consistent with the perturbative result $\beta = (2\pi^2 g)^{-1}$. The solid symbols are from L = 32 and L = 16, and may overestimate β for the reasons discussed above, especially in the large g limit when ℓ/L is large. These corrections appear to be small for $\ln g \lesssim -3$.

It has been shown that the leading logarithmic divergence in perturbation theory is destroyed by terms in the Hamiltonian which break time reversal symmetry, such as spin flip scattering [4] or a magnetic field [15]. Wegner [2] has considered models with complex hopping matrix elements and showed that $\beta(g) \approx -2/(4\pi^2 g)^2 + 0(g^{-4})$. To test this case we have performed calculations with $V_x = 1$ and $V_y^{jk} = \exp(i\theta_{jk})$ where θ_{jk} is a random phase. The results for purely off-diagonal disorder, W = 0, are shown in the insert in Fig. 2.

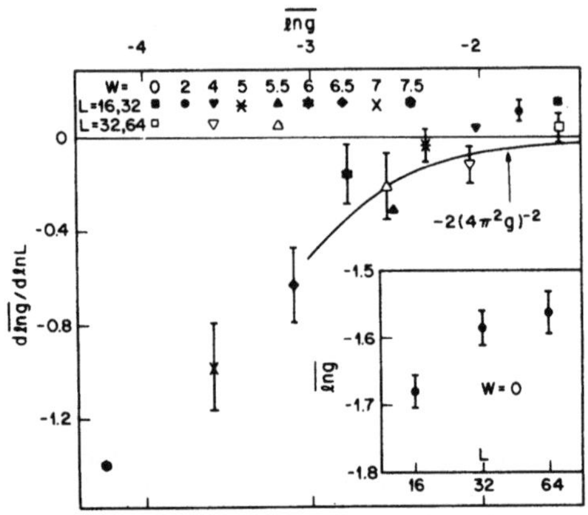

Fig. 2. Scaling function for the Anderson model with random complex hopping matrix element

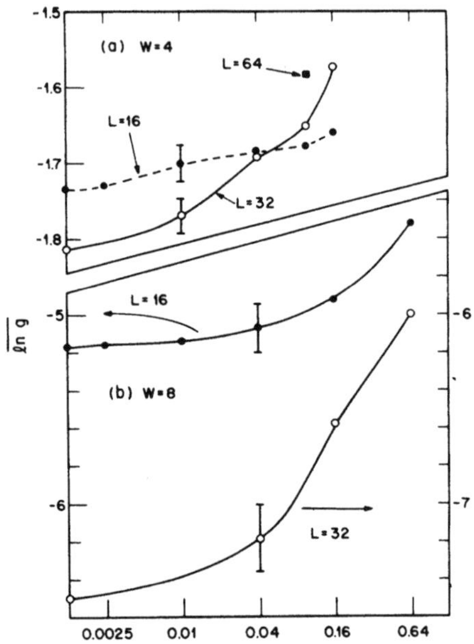

Fig. 3. Conductance vs magnetic field B measured in units of $\hbar c/ea^2$

The increase in g between L = 16 and 32 may be due to finite size effects. Comparison with the insert in Fig. 1 shows that the two cases with similar g show very different scaling behavior. A scaling function is constructed as before and shown in Fig. 2. Again the open symbols which are obtained from scale changes between L = 32 and 64 are consistent with the perturbative result. The uncertainties in our computation are such that we cannot hope to distinguish between a smooth nonzero β function and the existence of a line of fixed points, but it is clear that the complex hopping case is described by a very different β function than the real hopping case.

Lastly, we present some calculations of magnetoresistance. The hopping part of the tight binding model can be expressed in terms of momentum operators $\hat{p}_x$ and $\hat{p}_y$ so that $H_0 = V_x\exp(i\hat{p}_x) + V_y\exp(i\hat{p}_y)$ + c.c.. A vector potential is introudced by the prescription $\hat{p} \rightarrow \hat{p} - e\vec{A}$. For a uniform magnetic field B normal to the plane it is convenient to choose the gauge $A_y = Bx$, $A_x = 0$. Since A_y commutes with $\hat{p}_y$ this is equivalent to taking $V_x = 1$ and $V_y^{kj} = \exp iBj$ in (2), where B is measured in units of $\hbar c/a^2$ (a is the lattice spacing). The results for W = 4 are shown in Fig. 3a. For a given L we observe a negative magnetoresistance.

For a sufficiently large field we find that the logarithmic localization is destroyed. The conductance change for $B = 0.1$ as L varies from 16 to 64 is similar to the complex hopping case. This is again consistent with predictions based on the perturbation theory [15]. Interestingly the negative magnetoresistance persits into the exponentially localized regime as shown in Fig. 3b for $W = 8$. This suggests that the localization length *increases* with increasing magnetic field. We speculate that this occurs when the Landau radius $(eB/\hbar c)^{1/2}$ becomes comparable with the localization length. Apparently the time reversal breaking nature of the magnetic field delocalizes the electrons. It should be possible to understand the effect of a magnetic field as causing a cross-over from the real hopping case (Fig. 1) to the universality class of the complex hopping case (Fig. 2).

REFERENCES

[1] E. Abrahams, P. W. Anderson, D. C. Licciardello and T. V. Ramakrishnan, Phys. Rev. Lett. *42*, 673 (1979).

[2] F. Wegner, Z. Phys. *B 35*, 207 (1979).

[3] D. C. Licciardello and D. J. Thouless, Phys. Rev. Lett. *35*, 1475 (1975), and J. Phys. C *11*, 925 (1978).

[4] P. A. Lee and D. S. Fisher, Phys. Rev. Lett. *47*, 882 (1981).

[5] D. S. Fisher and P. A. Lee, Phys. Rev. *B23*, 6851 (1981).

[6] J. L. Pichard and G. Sarma, J. Phys. *C 14*, L127 (1981).

[7] A. MacKinnon and B. Kramer, Phys. Rev. Lett. *47*, 1546 (1981).

[8] J. Stein and V. Krey, Z. Physik *B37*, 13 (1980).

[9] E. N. Economou and C. M. Soukoulis, Phys. Rev. Lett. *46*, 618 (1981).

[10] D. J. Thouless, Phys. Rev. Lett. *47*, 972 (1981).

[11] D. Langreth and E. Abrahams, Phys. Rev. *B24*, 2978 (1981).

[12] R. Landauer, Phil. Mag. *21*, 863 (1970).

[13] P. A. Lee, Phys. Rev. Lett. *42*, 1492 (1979).

[14] P. A. Lee, J. Non. Crys. Sol. *35*, 21 (1980).

[15] B. L. Altshuler, D. Khmel'nitskii, A. I. Larkin and P. A. Lee, Phys. Rev. *B22*, 5142 (1980).

Numerical Study of DC Conductivity in Two-Dimensional Disordered Systems

S. Yoshino

Institute of Materials Science, University of Tsukuba
Sakura-mura, Ibaraki 305, Japan

Abstract

Electrical conductivity in two dimensions is investigated numerically by treating samples composed of up to 100×100 sites. The quantity which is proportional to the conductivity in the thermodynamic limit is introduced and computed using the eigenstates at the Fermi energy, and its dependence on a kind of complex frequency or the system size is examined. The difference between the computed and the classical (Drude-like) values is estimated and the logarithmic dependence is observed when the Fermi energy is not near the band center. If the Fermi energy is near the band center the situation is rather complicated. The effect of the system size and the accuracy of the computations are also discussed.

1. Introduction

Anderson localization of an electron in two dimensions is now an exciting problem. Concerning the conductivity at zero temperature in the weakly disordered systems with normal impurities and without mutual interactions or magnetic field, the following expression is obtained by the scaling theory [1], or summing up the totally crossed graphs in the two-particle kernel in the perturbation series [2], in addition to the usual ones deriving the relaxation time τ^0;

$$\sigma(\omega) = \sigma^0(\omega) + (e^2/2\pi^2\hbar)\ln(\omega\tau^0) + \ldots \qquad (1)$$

$\sigma^0(\omega)$ is the classical or the Drude-type conductivity and τ^0 is obtained by the Born approximation. Hereafter we mean by the 0-th approximation, the one neglecting the second term or more in (1), and denote it by the index 0 as a superscript. This approximation is good for not too small ω (or energy in general). The scaling theory insists that the conductivity decreases logarithmically and finally vanishes as ω goes to zero. The case $\omega \sim 0$ is also investigated by the self-consistent treatment [3]. Several numerical studies are also carried out along the scaling argument [4,5].

In the present numerical simulation we try to compute the dc conductivity in two dimensions by putting the scaling argument out of consideration and making ambiguous or sophisticated assumptions as little as possible. Then the numerical results are examined with main attention to the difference between the computed and the 0-th approximate values, which is to have the logarithmic dependence on energy in correspondence to (1). It is, of course, a quite difficult task, since the systems of finite size are treated, for which the conductivity cannot be rigorously defined. In addition the logarithmic dependence is quite ill-natured to examine, considering the noise of fluctuations inevitable in the treatment of finite size. τ^0 should be large enough, i.e., the degree of disorder should be weak, because (1) is the expansion in $1/\tau^0$. On the other hand τ^0 must not be too large for our purpose, since the logarithmic term comes to take smaller part in $\sigma(\omega)$ for larger τ^0 and to be difficult to distinguish. It is thus necessary to examine the cases of adequate degree of disorder and Fermi energy E_F.

This work is carried out as the extension of the work reported several years ago [6], where we treated two-dimensional disordered systems containing up to 100×100 sites, and visualized the localization by using such figures as Fig.1. The states shown in Fig.1 were given as the examples of the extended states, but such an intuitive demonstration can possibly be misunderstood. The existence of extended states was less confirmed than that of localized ones in [6]. It is shown in this paper that the states in Fig.1 are not necessarily extended, especially Fig.1(b) is the case of clear localization.

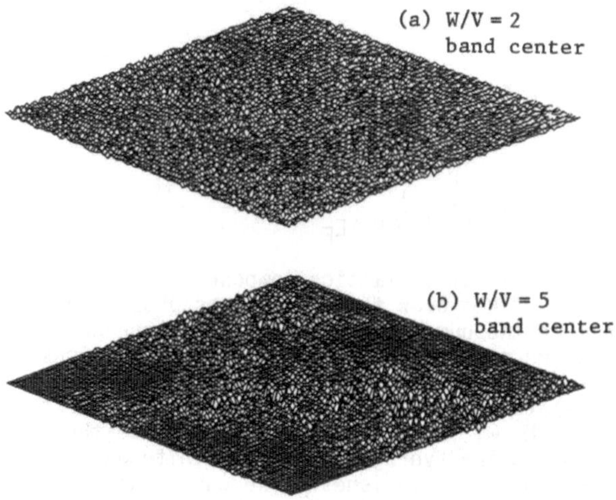

Fig.1 A bird's-eye view of amplitudes of an eigen state [6]

In this paper we mainly treat the Drude-like case in the 0-th approximation. The accuracy of the computations, which is important when large samples are treated, is also discussed. The details of this work will be published elsewhere.

2. Model and Computational Procedure

The model used in the present work is the traditional Anderson model on the two-dimensional square lattices. The Hamiltonian for a non-interacting electron is given by

$$H = -V \sum_{nn} |i\rangle\langle j| + \sum_i \varepsilon_i |i\rangle\langle i| , \tag{2}$$

where -V is the constant transfer integral between nearest neighbor sites, and the other transfers are zero. ε_i is a random site energy distributed uniformly in [-W/2, W/2]. A sample lattice is composed of $L\times L$ sites, where

L is varied as 40, 60 and 100. The periodic boundary condition is imposed in one direction (x-direction), and the rigid wall boundary condition is in the other (y-) direction. ε_i is generated by pseudorandom numbers.

To investigate the conductivity at zero temperature, we use the Kubo formula for the dc one;

$$\sigma_{dc}(T=0) = (2\pi\hbar e^2/\Omega)\mathrm{Tr}[v_\mu\delta(E_F-H)v_\mu\delta(E_F-H)], \qquad (\mu = x,y) \tag{3}$$

where v_μ is the velocity operator in the μ-direction, and Ω is the volume (area) of the system. Eq.(3) contains two δ-functions, which is the case of the system in the thermodynamic limit. Finite size samples are under consideration in the numerical study, so these δ-functions should be treated adequately. In the present work one δ-function is separated by introducing the density of states and the average on energy shell, and the other is replaced by the Lorentzian. We have then

$$\sigma_{dc} = 2e^2\langle v_\mu^2\rangle_{OS}\, g(E_F) \lim_{\eta\to 0}\lim_{L\to\infty} s_\mu(\eta,L) \tag{4}$$

where $g(E)$ is the density of states per a site, and $\langle\ \rangle_{OS}$ means the average over the states on energy shell. In (4) and hereafter we take for simplicity, $\hbar = V = \Omega/L^2 = 1$, and energies are measured from the band center. $s_\mu(\eta,L)$ is given by,

$$s_\mu(\eta,L) = \Big\langle \langle\alpha|v_\mu \mathrm{Im}(E_F-i\eta-H)^{-1}v_\mu|\alpha\rangle/\langle\alpha|v_\mu^2|\alpha\rangle \Big\rangle_{E_\alpha \doteqdot E_F} \tag{5}$$

where $|\alpha\rangle$ is the eigenstate of (2) on a sample lattice composed of $L\times L$ sites, and the average (arithmetic mean) is taken over the eigenstates whose energies are close to E_F. In (5) the term in the numerator is of significance and is divided by the diagonal element of v_μ^2 in order to diminish the fluctuations due to the finite size as much as possible. Strictly speaking, the average in (5) should be taken over such states as $|E_\alpha - E_F| < \delta$, where δ should go to zero after the thermodynamic limit, $L\to\infty$, is taken. As a consequence the number of eigenstates participating in the average becomes infinite as $L\to\infty$, and then the width of the distribution of s_μ vanishes if we can regard each quantity in $\langle\ \rangle$ as random. Now we are using the finite number of eigenstates, and thus should estimate the reliable value of (5) by $m\pm d/\sqrt{n}$, where m and d are the mean and the root-mean-square deviation when averaged over n eigenstates. The variation $\pm d/\sqrt{n}$ is indicated by error bars in all figures in the next section, although it is not always drawn because it is often smaller than the size of the mark.

In the 0-th approximation the values of various physical quantities are expressed in terms of the complete elliptic integral with a complex argument, because we are using the tight-binding model, not the free-electron one as the unperturbed part. They are, however, often treated in this paper as the same as those in the free-electron model. The difference is rather small in most cases, and an exception to it is that when E_F is near the band center, where the van Hove singularity appears in the tight-binding model while none in the free-electron one.

The following equation holds for $s_\mu(\eta,L\to\infty)$ in the weakly disordered case corresponding to (1);

$$s_\mu(\eta,\infty) = s^0(\eta) + [1/4\pi^2\langle v_\mu^2\rangle_{OS}^0\, g^0(E_F)]\ \ln(\eta\tau^0). \tag{6}$$

$s^0(\eta) \equiv s_\mu^0(\eta,\infty)$ is also expressed by the complete elliptic integral, but approximately written by the Drude-like expression, namely,

$$s^0(\eta) \simeq (1/\tau^0 + \eta)^{-1}. \qquad (7)$$

2.1 Rough Estimate of the Logarithmic Term

As is mentioned before, the degree of disorder, W, should be small enough but not be too small to see the logarithmic term in (1) or (6). To satisfy this restriction the case of $W = 2$ is treated and E_F is varied as ±0.5 and ±1.5 as far as this paper is concerned. $s^0(0) \simeq \tau^0$ can be estimated by the following relation;

$$1/2\tau^0 = \pi(W^2/12)g^0(E_F). \qquad (8)$$

When W=2 and E_F is not near the band edges, then we have $s^0(0) \simeq \tau^0 \sim 4$ since $g^0(E_F) \sim 1/8$.

Two characteristic energies can be considered corresponding to the finite size. One is the mean energy level spacing, Δ_S. The other, Δ_T , is related to the boundary effect for a moving electron and is given by the inverse of the time for an electron to travel across the sample. They are thus given by,

$$\Delta_S = 1/L^2 g^0(E_F) \ , \qquad \Delta_T = (\langle v_\mu^2 \rangle_{OS}^0)^{1/2}/L \ , \qquad (9)$$

where the use of the 0-th approximate values is not serious. Now we have for $L = 40 \sim 100$ by using $\langle v_\mu^2 \rangle_{OS}^0 \sim 2$,

$$-[1/4\pi^2 \langle v_\mu^2 \rangle_{OS}^0 \, g^0(E_F)] \ln(\Delta_S \tau^0) \sim 0.4 \sim 0.6 \ ,$$

$$-[\qquad \text{do.} \qquad] \ln(\Delta_T \tau^0) \sim 0.2 \sim 0.3 \ .$$

Δ_S and Δ_T can be regarded as the lower bound of η to make the computation of significance. Consequently we can have the logarithmic term in (6) with the magnitude which is about 10% or less of the leading term, $s^0(\eta)$. This ratio is small but not too small to distinguish in the noise of fluctuations. The relation $\Delta_S \ll \Delta_T$ must be satisfied because we are treating two dimensions, not one dimension.

2.2 Expected Behavior of $s_\mu(\eta,L)$ for the Metallic Region

In this subsection the behavior of s_μ is discussed for the system, where such singular term as the logarithmic one do not exist and $\sigma(\omega)$ is well-behaved when $\omega \to 0$. In the 0-th approximation the square modulus of the matrix element of the velocity, $|\langle\alpha|v_\mu|\beta\rangle|^2$, can be approximated by the Lorentzian with half-value width $2/\tau^0$, as a function of the energy difference, $E_\alpha - E_\beta$. This behavior is, however, true only when the thermodynamic limit is taken, and for the finite size system some modification is to be taken into account.

In the rigid wall boundary (y-) direction, $|\langle\alpha|v_y|\beta\rangle|^2$ is written using the element of the position operator as

$$|\langle\alpha|v_y|\beta\rangle|^2 = (E_\alpha - E_\beta)^2 \, |\langle\alpha|y|\beta\rangle|^2. \qquad (10)$$

By this relation $s_y(\eta,L)$ is expected to behave as in Fig.2(a); s_y starts from zero as $\langle\Delta y^2\rangle_{OS} \times \eta \sim O(L^2) \times \eta$ for sufficiently small η, and takes the maximum at $\eta \sim \eta_M$, where $\eta_M^2 \simeq \langle v_y^2\rangle_{OS}/\langle\Delta y^2\rangle_{OS}$. Since $\langle\Delta y^2\rangle_{OS} \simeq L^2/12$,

$$\eta_M \simeq (12\langle v_y^2\rangle_{OS}^0)^{1/2}/L. \qquad (11)$$

If L becomes larger and larger, then η_M goes to zero and $s_y(\eta,L)$ approaches to the one in the thermodynamic limit, i.e., $s_y(\eta,\infty)$ is finite when $\eta \to 0$.

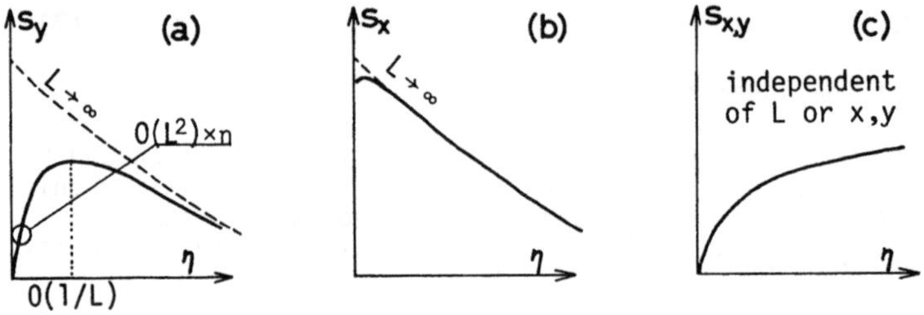

Fig.2 Expected behavior of $s_\mu(\eta,L)$ in three cases; (a) metallic case and μ in the rigid wall direction, (b) metallic and μ is in the periodic direction, and (c) insulating case.

In the periodic boundary (x-) direction, the relation like (10) does not hold, because the position operator x cannot be defined. In the finite size system the eigenstate is always expressed by a real amplitude at every site and degenerate states do not appear except in the accidental case, so the diagonal element of v_x vanishes. This fact results in the correction term by the finite size, and including this term we have as a corrected one of s_x,

$$s_x^c(\eta,L) = s_x(\eta,L) + [\tau^0/\pi g^0(E_F)]/\eta L^2. \tag{12}$$

This correction vanishes when the limits are taken in the correct order (first $L\to\infty$, and then $\eta \to 0$).

Now the logarithmic term neglected in the above discussion is taken into account. In the rigid wall direction case this term has little effect on the above argument for the system as large as $L \sim 100$. In contrast, in the periodic boundary case the correction given by (12) may be redundant, since the matrix element $|\langle\alpha|v_x|\beta\rangle|^2$ may continuously decrease to zero according to the scaling theory, while above argument assumes that the matrix element vanishes only at one point, $E_\alpha - E_\beta = 0$, and has a finite value in the neighborhood, which gives the finite dc conductivity. If we calculate the matrix element of v_x itself and plot it vs $E_\alpha - E_\beta$, which is not given in this paper, the results agree with the latter (only one point vanishing), so the correction is taken into account as (12) in this work.

2.3 $s_\mu(\eta,L)$ for the Insulating Region

In this case the situation is simpler than in the former case. $s_\mu(\eta,L)$ is expected to behave as in Fig.2(c), independent of the size L or the direction (μ = x or y), reflecting the electron localization at E_F. s_μ, of course, decreases when η comes to be very large, so it must have the maximum at large η. For the position of this maximum, the same argument is available as in the case of the rigid wall direction in the metallic region. In the present case, $\langle\Delta y^2\rangle_{0S} \sim O(\ell^2)$ and thus the position is $\sim O(1/\ell)$, which is independent of L and has the order unity, where ℓ is the localization length.

3. Results and Discussions

In all figures in this section, the number of eigenstates used in the evaluation of (5) is as follows; when one sample is given, ten eigenstates are com-

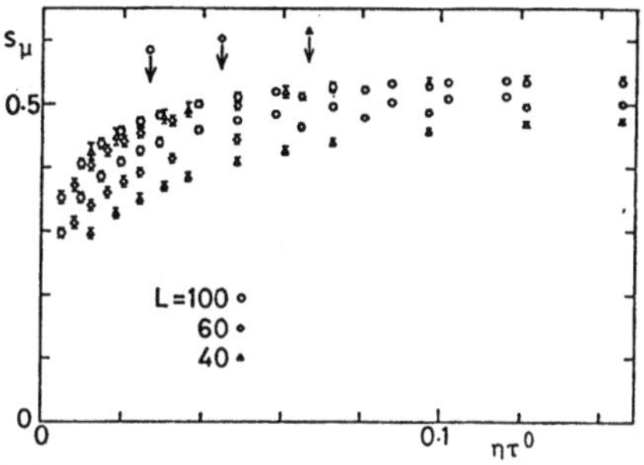

(a) W/V = 5
$E_F/V = 0$

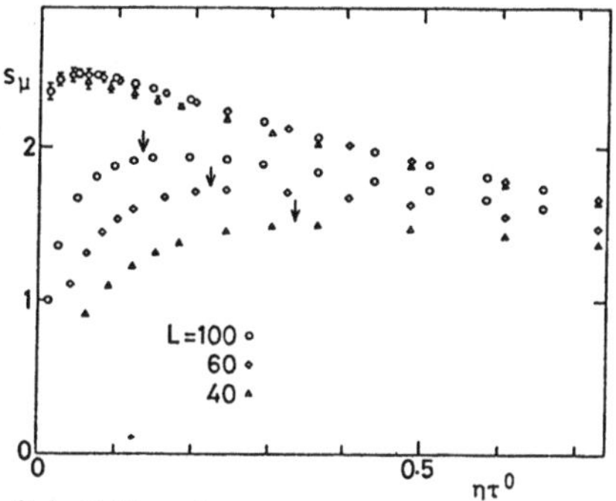

(b) W/V = 2
$E_F/V = \pm 0.5$

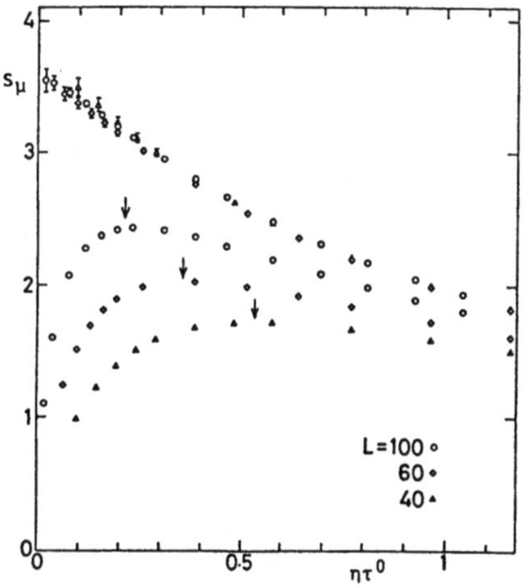

(c) W/V = 2
$E_F/V = \pm 1.5$

Fig.3 $s_\mu(\eta,L)$ vs η. In all figures the upper three are s_x and the lower three are s_y. The arrows indicate the positions of η_M given by (11). Note that the scale of the ordinate in (a) is different from that of (b) or (c)

puted whose energies are closest to E_F, and eight or sixteen samples are treated for one set of W, E_F and L. The case of sixteen samples is that where W = 2, $E_F = \pm 1.5$ and L = 40 or 60, and eight samples in the other cases. Thus the average in (5) is taken over 80 or 160 eigenstates.

The results of $s_\mu(\eta,L)$ are shown in Fig.3 in three cases. In connection to Fig.1 the case where W = 5 and $E_F = 0$ is also given in Fig.3(a). In this case s_x and s_y both decrease when η is decreased for any value of L, although the tendency to go to zero is not clear because of the lack of data near zero. This is the characteristic of insulating region, so we can conclude that the states at E_F are well localized. It is also noticed that s_x is almost independent of L, while s_y is a little smaller than s_x and slightly increases as L increases and then approaches to s_x. This phenomenon is to be ascribed to the fact that the spatial extent of the localized wavefunction is comparable with the sample size, as is also seen in Fig.1(b). The localization center is not well defined and consists of the islands of centers connected with one another, while it is well defined in the case of larger disorder [6].

Figures 3(b) and (c) are both in the case of $W=2$, and the difference is whether E_F is near the band center (b), or not (c). The eigenenergy (very close to zero) of the state shown in Fig.1(a) is not the same as E_F ($=\pm 0.5$) in Fig.3(b), but these two cases are similar to each other in the sense that both energies are near the band center. The similar figure to Fig.3(b) is obtained if the data for $E_F=0$ are plotted, although it is not given in this paper. If we neglect the region near $\eta \sim 0$, then s_x and s_y both show the expected behavior for metallic region, i.e., s_y has the maximum near η_M and s_x monotonously increases when η decreases. More careful investigation of Fig.3(b) shows that the maximums of s_y are slightly shifted from the arrows in addtion to the decrease of s_x near $\eta \sim 0$, which is not the case in Fig.3(c).

To clarify the term corresponding to the logarithmic one in (6) and its dependence on η, the following plot is done and given in Fig.4; we introduce the correction given by (12) and regard $s_x^c(\eta,L)$ as being nearly equal to $s_x(\eta,\infty)$. Then $\Delta s_x(\eta,L) \equiv s_x^c(\eta,L) - s^0(\eta)$ is plotted vs $\log(\eta\tau^0)$ in considera-tion of (6), where $s_x^c(\eta,L)$ is now used instead of $s_x(\eta,\infty)$. Comparing Figs. 4(a) (near the band center) and (b) (not near the band center), the following facts are noticed.

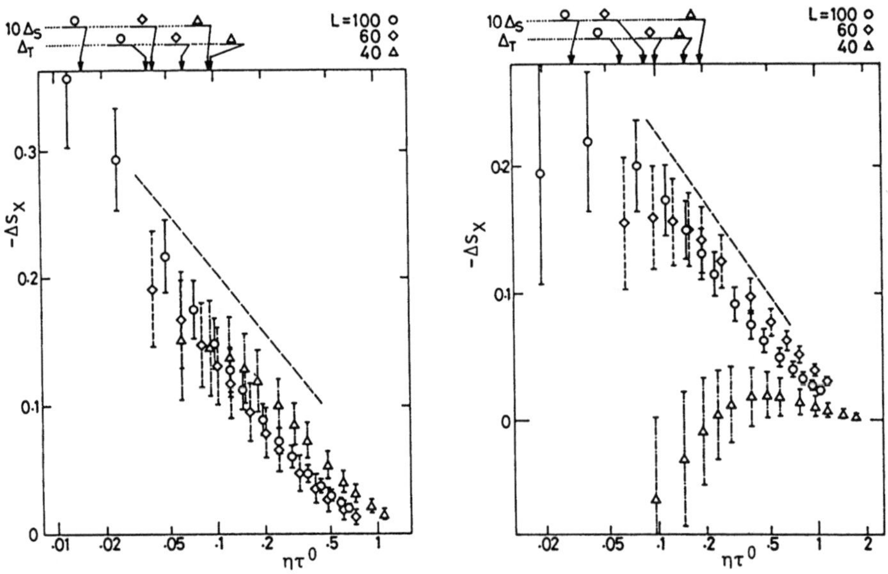

(a) $W/V = 2$
$E_F/V = \pm 0.5$

(b) $W/V = 2$
$E_F/V = \pm 1.5$

Fig.4 $\Delta s_x(\eta,L)$ vs $\log(\eta\tau^0)$. The broken straight line indicates the gradient of the logarithmic term in (6). Note that $-\Delta s_x$ is in the ordinate, so Δs_x decreases if a plot goes upwards, and v.v.

(1) When the system size is varied, the variation of Δs_x is smaller in the case (a) than in (b), which may be ascribed to the ratio of Δ_S to Δ_T. $10\Delta_S$ and Δ_T are indicated at the top of the figures. The relation $\Delta_S \ll \Delta_T$ holds better in (a) than in (b). That is to say, the nearer the system comes to one dimension, then the more deviation Δs_x shows. The direction of this

deviation is not necessarily downwards in Fig.4 (increase of Δs_X), considering the case not reported in this paper. In every case Δs_X shows the change of the tendency of its behavior near the point where η equals to $10\Delta_S$ or more; i.e., when η is decreased from $\sim 1/\tau^0$, first Δs_X decreases and then shows smaller decrease (a) or increase (b) after this point is passed.

(2) The gradient of a plot in Fig.4 is compared with the broken line. Concentrate our attention on the circles, the case of $L = 100$, then it is noticed that in the case (b) they agree with each other in the region where η is greater than the turning point discussed above and less than $\sim 1/\tau^0$, while they do not in the case (a). The data for $L = 40$ or 60 are less significant for this purpose, since the gradient is forced to be less steep by the greater turning point.

Now we mean by "metallic", the case where the expansion like (1) holds, i.e., the conductivity is expressed by the expansion with the Born approximate value as the leading term. Then the system with E_F not near the band center is metallic, while the non-metallic region appears if E_F comes to be near the band center. To make the characteristics of this region clearer more detailed study is needed, and the logarithmic van Hove singularity at the band center should be, of course, taken into account.

4. Accuracy of Computations

In the present work we have used not only eigenvalues but also eigenvectors of the matrix form of (2), so we should pay attention to the accuracy of eigenvectors to confirm our results. In this section we give a brief comment on the errors of the computed eigenvectors and $s_\mu(\eta,L)$ in addition.

Let E and $|\alpha\rangle$ be the approximate (computed) eigenvalue and eigenvector, and $|i\rangle$ be the exact eigenvector corresponding to the exact eigenvalue E_i. For arbitrary scalar Δ, we have

$$\begin{aligned}\langle\alpha|(E-H)^2-\Delta^2|\alpha\rangle &= \sum_i |\langle\alpha|i\rangle|^2\,[(E-E_i)^2-\Delta^2] \\ &\geqq -\Delta^2 \sum_{i\in C} |\langle\alpha|i\rangle|^2, \end{aligned} \tag{13}$$

where C denotes the set of eigenvectors whose eigenvalues lie in $[E-\Delta, E+\Delta]$. Then it is easy to obtain

$$1 - \sum_{i\in C} |\langle\alpha|i\rangle|^2 \leqq \|(E-H)|\alpha\rangle\|^2/\Delta^2. \tag{14}$$

If there exist only one eigenvector in C, the summation in (14) is not necessary and the r.h.s. of (14) gives the upper bound of the error of the eigenvector $|\alpha\rangle$. In our case Δ is to be replaced by the energy difference between adjacent levels which is of the same order as Δ_S. So we can insist that one eigenstate is obtained quite accurately and separated from the others with close eigenenergies, if the following condition is satisfied,

$$\|(E-H)|\alpha\rangle\| \ll \Delta_S. \tag{15}$$

In the present work, $\Delta_S \sim 10^{-3}$ for $L = 100$, and $\|(E-H)|\alpha\rangle\| \sim 10^{-5}$ in the worst case and $10^{-8} \sim 10^{-12}$ in general, so this condition is well satisfied. The term "worst case" means that we have truncated the iteration at the stage where the worst value of them reaches 10^{-5}. We can, of course, improve the accuracy by repeating the iteration more times.

The present computation, where the double precision operation with 16 significant digits is used, shows that when $L=100$ the value of $\|(E-H)|\alpha\rangle\|$ can be reduced to no less than 10^{-12} for any state if the iteration is sufficiently repeated. In other words the numerical errors are piled up to the order of 10^{-12} or more in the computation of eigenvectors. This magnitude is quite reasonable and desirable. Let ε be the rounding error involved in one arithmetic operation. Consider a numerical result obtained after N operations, then the rounding error in this result comes to be of order $\sqrt{N}\varepsilon$, taking account of the cancellation of positive and negative errors. In our case $N \sim L^4$, $L=100$ and $\varepsilon \sim 10^{-16}$, so the errors of computed eigenvectors should be of order 10^{-12}, which coincides with that in the actual computation. This fact shows that the errors in eigenvectors mainly come from the rounding errors and do not suffer from the numerical instability or the latent defect of the procedure which often crops out when the amount of computations becomes larger. Now we estimate the largest value of L for which satisfactorily accurate eigenvectors can be obtained. Using $\Delta_S = 1/L^2 g^0(E_F)$ and $\|(E-H)|\alpha\rangle\| \sim L^2\varepsilon$ we have from (15) that $L \ll 10^4$, if the double precision operations are used. Fortunately we are far away from this value. If the single precision operations are used, where $\varepsilon \sim 10^{-7}$, then $L \ll 100$, which does not answer our purpose.

Concerning the accuracy of s_μ, the situation is rather good. Most part of the computation is to calculate the inverse of the matrix $(E_F-H)^2+\eta^2$ (not the inverse itself but a vector multiplied by the inverse [6]), which is the same procedure as is already used in the case of eigenvectors, where ill-natured trouble has not come. The amount of arithmetic operations needed for s_μ is greater than that for eigenvectors but of the same order; $L^4/2$ for eigenvectors and $2L^4$ for s_μ. So only errors of order 10^{-12} are brought in s_μ, which is not serious.

5. Summary

From the numerical results of s_μ and the difference of s_x from the Drude-like expression, we have Fig.5 as the diagram to show the metallic and insulating regions. In Fig.5 the term "metallic in some sense" means that the expansion as (1) holds to some extent, while in the region "localized?" the deviation from this expansion is recognized and the numerical results draw the clear distinction between them. In the latter region the van Hove singularity in the case of no disorder should be considered.

E_F	0	3∿4	6∿7 W/V
not near the band center	metallic in some sense	exponential localization	
near the band center	localized?	*not* well defined loc. center	well defined loc. center

Fig.5 Diagram for metallic or insulating character

Finite size effects should be, of course, taken into account in the numerical study, but at the same time the accuracy and the stability of the computations should be examined when the amount of computations becomes large. In the present work we can estimate the errors of the results as in §4.

References

1 E. Abrahams, P. W. Anderson, D. C. Licciardello and T. V. Ramakrishnan: Phys. Rev. Lett. 42, 673 (1979).
2 J. S. Langer and T. Neal: Phys. Rev. Lett. 16, 984 (1966).
3 P. Wölfle: in this symposium.
4 P. A. Lee: in this symposium.
5 A. MacKinnon: in this symposium.
6 S. Yoshino and M. Okazaki: J. Phys. Soc. Jpn. 43, 415 (1977).

Interaction Effects in Disordered Systems

P.A. Lee

Bell Laboratories
Murray Hill, NJ 07974, USA

ABSTRACT

Recent developments in the theory of disordered interacting systems are reviewed. In the weakly interacting and weak disorder limit, important corrections are found for the electron-electron scattering rate, tunneling density of states, conductivity and magnetoconductivity.

I. INTRODUCTION

The study of the interacting electron gas has a long history. Early studies using perturbation theory in the unscreened Coulomb interaction led to strong singularity near the Fermi surface. It was then realized that a proper calculation taking screening into account removed all the singularities. (See PINES and NOZIERES [1]). These results can be fitted into the general framework of Landau's Fermi liquid theory which concludes that the effects of interaction can be represented by the introduction of a number of Fermi liquid parameters which describe the renormalization of physical quantities such as specific heat and magnetic susceptibility. While these renormalization can be large, they are finite. In the early sixties these studies were extended to the vicinity of the ferromagnetic transition. It was found that low lying spin fluctuations (paramagnons) lead to strong renormalization of physical quantities, and in addition, introduce singularities near the Fermi surface. However these singularities are typically weak, such as a $T^3 \ln T$ term in the specific heat. The general feeling was that impurity scattering would not lead to any essential modification of the Fermi liquid theory (see BETBEDER-MATIBET and NOZIERES [2]). Thus it came as quite a surprise when it was shown by ALTSHULER and ARONOV [3] that interactions in a disordered Fermi liquid lead to strong singularities near the Fermi level. For example a singularity of the form $(\omega - E_F)^{1/2}$ is predicted for the tunneling density of states and $T^{1/2}$ and $T^{3/2}$ low temperature corrections are predicted for the conductivity and the specific heat respectively. Many of these results were obtained independently by FUKUYAMA [4]. Interestingly some of these effects were already anticipated by BRINKMAN and ENGELSBERG [5], who showed that the diffusive nature of spin fluctuations in a disordered system lead to stronger singularities. In particular the $T^3 \ln T$ term in the specific heat becomes $T^{3/2}$. The ALTSHULER-ARONOV effect can be understood if we replace spin diffusion in BRINKMAN-ENGELSBERG by density diffusion. The irony is that perhaps the Fermi liquid theory was so successful and so entrenched in our thinking that our understanding of the disordered Fermi liquid may have been delayed by more than a decade.

II. SELF ENERGY CORRECTIONS AND LIFETIME IN DISORDERED FERMI LIQUID

1. Model Calculation of the Density of States

ALTSHULER and ARONOV [3] treated the disordered Fermi liquid problem by performing a perturbation theory to lowest order in the interaction strength. The disorder is treated by the conventional diagrammatic technique (ABRIKOSOV, GORKOV and DZYALOSHINSKII [6]) which should be valid in the limit $k_F\ell \gg 1$. In this limit crossed impurity lines are considered higher order in $(k_F\ell)^{-1}$ and are ignored. Thus the localization effects discussed recently based on maximally crossed diagrams [7] are specifically excluded. For simplicity we shall first discuss a model problem in which the electrons interact via a static interaction v(q)

$$H_1 = \tfrac{1}{2}\sum_q v(q)\rho_q^+\rho_q \tag{1}$$

where $\rho_q = \sum_k a_{k+q}^+ a_k$. We shall return to discuss the complications of the dynamically screened Coulomb interaction later.

The important ingredient of the interaction theory is that the interaction vertex is dressed by the impurity scattering as shown in Fig. (1). A straight forward calculation shows that for small q and ω_m, the vertex correction is given by

$$\Gamma(q,\omega_m,\epsilon_n) = \begin{cases} (|\omega_m|+Dq^2)^{-1}\tau^{-1} & \text{if } \epsilon_n(\epsilon_n-\omega_m) < 0 \\ 1 & \text{otherwise} \end{cases}, \tag{2}$$

where $\omega_m = 2\pi mkT$ and $\epsilon_n = \pi(2n+1)kT$ are the Matsubara frequencies. Note that when the electron lines have frequencies with opposite signs, the vertex correction has the diffusion form and become singular in the limit of small q and ω_m. The relation with density diffusion is best illustrated by calculating the density correlation function $D(q,\omega_m)$

$$D(q,\omega) = i\int_0^\infty dt dx \overline{\langle[\rho(\vec{x},t),\rho(0,0)]\rangle} e^{-iq\vec{x}} e^{i\omega t}$$

$$= \frac{dn}{d\mu}\frac{Dq^2}{-i\omega+Dq^2} \quad . \tag{3}$$

Eq. (3) can be derived from more general consideration by making the assumption that density fluctuations satisfy the diffusion equation (FOERSTER [8]).

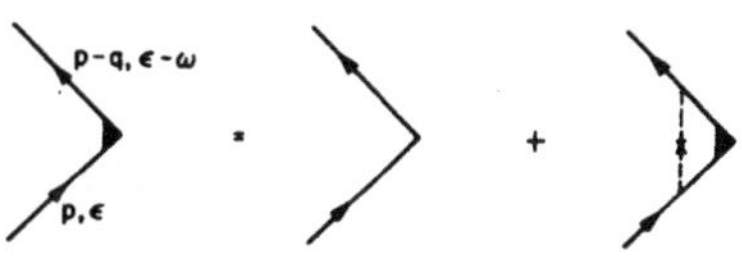

Fig. 1

Fig. 2

To lowest order in the coupling constant, the correction to the Green's function in the momentum representation is given by the diagram shown in Fig. 2.

The physical quantity which is independent of representation is the single electron density of states

$$\delta N(\omega) = \frac{1}{\pi}\sum_{k} \mathrm{Im}G(k,\omega) \quad . \tag{4}$$

This was first calculated by ALTSHULER and ARONOV [4], who showed that the diffusion pole associated with the two vertex corrections in Fig. (2) leads to unexpected singular behavior in the density of states, such as an $\omega^{1/2}$ behavior in three dimensions. We emphasize that the singularities are unrelated to the long range nature of the Coulomb interaction. In fact starting with the short range model interaction we see that the vertex correction generates an effective long range and retarded coupling. This expresses the physical idea that since the electron motion is diffusive, the electrons spend a longer time in a given region in space relative to the plane wave states and their interaction is enhanced.

The diagrammatic treatment is carried out in the basis of momentum eigenstates. The plane wave is very far from being an eigenstate of the disordered Hamiltonian. This is why in a perturbation treatment of the interaction, large vertex corrections are necessary. Furthermore quantities like self energy and decay rate are meaningless in the k representations because the spectral representation contains a large width given by the elastic scattering rate τ^{-1}. A more natural basis for the perturbation expansion is the set of exact eigenstates ψ_m with eigenvalue E_m of the noninteracting Hamiltonian. (ABRAHAMS et al. [9], MALDAGUE [10]). In this basis the single particle Green's function is given by

$$G_{mm}(\omega) = \langle m|(\omega-H)^{-1}|m\rangle \tag{5}$$

and the effect of interaction is included in the self energy correction $\Sigma_m(\omega)$ so that

$$G_{mm}(\omega) = (\omega-E_m-\Sigma_m(\omega))^{-1} \quad . \tag{6}$$

We keep only the diagonal terms in the self energy because we shall find that it is enhanced by wave-function correlation. Writing $\Sigma_m = \Delta_m + i\,\Gamma_m$, we perform the standard expansion

$$\Delta_m(\omega) = \Delta_m(\tilde{E}_m) + (\omega-\tilde{E}_m)\,\frac{\partial\Delta_m}{\partial\omega} \tag{7}$$

where

$$\tilde{E}_m = E_m + \Delta_m(\tilde{E}_m) \quad . \tag{8}$$

We obtain

$$G_{mm}(\omega) = Z/[(\omega-\tilde{E}_m)-i\gamma_m] \tag{9}$$

where

$$Z^{-1} = [1-\partial\Delta_m/\partial\omega]_{\omega=\tilde{E}_m} \tag{10}$$

and

$$\gamma_m = Z\,\Gamma_m(\omega=\tilde{E}_m) \quad . \tag{11}$$

In (11), $\tilde{E}_m$ is interpreted as the quasi-particle energy, γ_m the decay rate of the quasi-particle and Z is the fractional weight of the quasi-particle excitation.

To discuss the average energy shift and the average decay rate we have to study the impurity average of the self energy for a fixed E_m, i.e.

$$\tilde{\Sigma}_E(\omega) = \frac{1}{N_0} \overline{\sum_m \delta(E-E_m)\Sigma_m(\omega)} \tag{12}$$

where N_Q is the one-spin density of states and the bar denotes impurity averaging. Writing $\tilde{\Sigma}_E = \tilde{\Delta}_E + i\tilde{\Gamma}_E$, we see that the average density of quasi-particle states is given by

$$N_Q = \frac{N_0}{1+\partial\tilde{\Delta}_E/\partial E} \tag{13}$$

and the average quasi-particle decay rate is given by

$$\tilde{\gamma}_E = \tilde{Z\Gamma} \approx \tilde{\Gamma}_E \tag{14}$$

where $\tilde{\Gamma}_E = (1/N_0)\overline{\sum_m \delta(E-E_m)\Gamma_m(\omega=\bar{E}_m)}$.

The model problem of electrons interacting via a static interaction $v(\vec{r})$ is particularly simple in that $Z = 1$ and $\hat{\gamma}_E = 0$. In the Hartree-Fock approximation, the four Fermion term in (1) is factorized. Let us focus our attention on the exchange term; the Hartree term can be shown to be small if the potential v(r) has a range larger than the interparticle spacing. We shall see that the diagonal exchange term is enhanced because of wave function correlation, so that we write $H_{int} = \sum_m \Sigma_m a_m^+ a_m$ where

$$\Sigma_m = - \sum_{n \text{ occupied}} \int d\vec{r}d\vec{r}' \, \psi_m^*(\vec{r})\psi_n^*(\vec{r}')\psi_m(\vec{r}')\psi_n(\vec{r})v(\vec{r}-\vec{r}') \ . \tag{15}$$

Suppose we insert a particle at energy E. Its energy will be shifted on the average by the amount

$$\begin{aligned}\tilde{\Sigma}_m &= \frac{1}{N_0} \sum_m \overline{\delta(E-E_m)\Sigma_m} \\ &= - \int_{-\infty}^{0} dE' \, F(E,E';\vec{r},\vec{r}')v(\vec{r}-\vec{r}')\end{aligned} \tag{16}$$

where

$$F(E,E';\vec{r},\vec{r}') = \sum_{m,n} \overline{\delta(E-E_m)\delta(E'-E_n)\psi_m^*(\vec{r})\psi_n^*(\vec{r}')\psi_m(\vec{r}')\psi_n(\vec{r})} \ . \tag{17}$$

In (17) we need to know the average of a product of four wavefunctions. Since we do not have an explicit solution of the impurity problem, we seem to be faced with a hopeless task. Fortunately we notice that the combination required in (17) is precisely that which enters in a density-density correlation function defined as

$$A(q,\omega) = \int_{-\infty}^{\infty} dt d\vec{r} d\vec{r}' \, e^{i\omega t} e^{i\vec{q}\cdot(\vec{r}-\vec{r}')} <[\rho(\vec{r},t),\rho(\vec{r}',0)]> \ . \tag{18}$$

This is the spectral function for the density-density response function given in (3). Again, under the assumption of density diffusion alone, we obtain

$$A(q,\omega) = \frac{dn}{d\mu} \, \mathrm{Im} \left[\frac{Dq^2}{-i\omega+Dq^2}\right] \ . \tag{19}$$

To make the connection with (17) we expand the operator ρ in (19) in terms of the exact eigenstates. Restricting ourselves to $T = 0$ for simplicity, for $\omega > 0$ only one ordering of the commutator is non vanishing and (18) becomes

$$A(q,\omega) = \sum_{m,n}{}' \int d\vec{r} d\vec{r}' \, e^{i\vec{q}\cdot(\vec{r}-\vec{r}')} \, \overline{\psi_m(\vec{r})\psi_n^*(\vec{r})\psi_m(\vec{r}')\psi_n(\vec{r}')\delta(E_n-E_m-\omega)} \tag{20}$$

where the sum is restricted to n occupied and m unoccupied. We convert the sum into an energy integration and compare with (17). We have

$$A(q,\omega) = \int_0^{\infty} dE \int_{-\infty}^{0} dE' F(E,E',\vec{r},\vec{r}') e^{i\vec{q}\cdot(\vec{r}-\vec{r}')} \delta(E-E'-\omega) \tag{21}$$

We expect $F(E,E';\vec{r},\vec{r}')$ to be dependent on $E - E'$ and $\vec{r} - \vec{r}'$ (the latter because we have translational invariance after impurity averaging). Then (21) becomes

$$A(q,\omega) = \omega \int F(\omega,\vec{r}) e^{i\vec{q}\cdot\vec{r}} d\vec{r}' \tag{22}$$

Comparing with (19), we have

$$\int F(\omega,\vec{r}) e^{i\vec{q}\cdot\vec{r}} \, d\vec{r} = \frac{dn}{d\mu} \frac{Dq^2}{\omega^2+(Dq^2)^2} \tag{23}$$

Eq. (23) is remarkable because it is divergent in the limit q, $\omega \to 0$. Going back to the definition $F(\omega,\vec{r})$ we see that the seemingly innocent assumption of diffusion implies a certain correlation between wavefunction overlap for eigenstates that are nearby in energy. We are now in a position to calculate the self energy using (16) and (23)

$$\tilde{\Sigma}_E = \frac{dn}{d\mu} \int_{-\infty}^{0} dE' \int d\vec{q} \frac{Dq^2}{(E-E')^2+(Dq^2)^2} \, v(\vec{q}) \tag{24}$$

According to (24) the exchange interaction between the added electron with energy E that the electrons in the Fermi sea depends strongly on the energy separation. As a result the self energy is also dependent on E. This will give rise to a change in the density of states, given by

$$\frac{\delta N}{N_0} \approx \delta\tilde{\Sigma}_E/\partial E = \int \frac{d\vec{q}}{(2\pi)^d} \frac{Dq^2 v(q)(dn/d\mu)}{E^2+(Dq^2)^2} \tag{25}$$

The integrand in (25) is singular in the limit q, $E \to 0$. We can therefore replace $v(q)$ by $v(0)$. Just from power counting we immediately see that δN is logarithmically divergent in 2D and goes like $\sqrt{E}$ in 3D. Eq. (25) agrees with results based on summation of diagrams [4].

In an electron gas it is necessary to introduce a dynamically screened Coulomb interaction. The polarization function that enters into screening is affected by impurities in the way shown in (2). In 3D we have

$$V_{C3}(q,\omega_n) = \frac{4\pi e^2}{q^2} \frac{|\omega_n|+Dq^2}{|\omega_n|+DK_3^2} \tag{26}$$

where $K_3^2 = 4\pi e^2(dn/d\mu)$. In strictly 2D (such as the Si-MOSFET inversion layer), we have

$$V_{C2}(q,\omega_n) = \frac{2\pi e^2}{q} \frac{|\omega_n|+Dq^2}{|\omega_n|+DqK_2} \tag{27}$$

where $K_2 = 2\pi e^2(dn/d\mu)$. It is interesting to note that in the static and long wave length limit ($\omega=0$, $q\to 0$) the dimensionless coupling constant $V_C(dn/d\mu) = 1$. The bare coupling constant e^2 is canceled in the problem.

2. Dynamic Screening and Electron Lifetime

The dynamic aspect of the effective interaction introduces a number of complications in that $\tilde{\Sigma}$ is now complex and depends on ω. The quasi-particle now acquire a lifetime due to the electron electron interaction contained in the elastic (imaginary) part of the dynamically screened interaction. If we compute $\tilde{\Sigma}$ using (15) with a complex V given by (26) and (27), we obtain a decay rate $\tilde{\Gamma}_E$ which varies as $E^{d/2}$. This result in 3D was first obtained by SCHMID [11]. Note that while the decay rate is enhanced compared with the standard E^2 result, it is still small compared with E so that quasi-particle concept is still valid. In 2D a further complication arises because of the existence of logarithmic singularities. The simple argument described here is not sufficient to yield the full ω dependence. ABRAHAMS et al. [9] showed that it is possible to rewrite (14) in terms of Green's functions and then perform the averaging using standard diagrammatic techniques in the momentum representation. The result is that at $T = 0$;

$$\tilde{\Gamma}_E^{2D} = \frac{e^2 K_2}{4}\frac{1}{\epsilon}\left\{\omega + (E-\omega)\ln\left|\frac{E-\omega}{\epsilon}\right|\right\} \tag{28}$$

while at finite T,

$$\tilde{\Gamma}^{2D} = e^2 K_2 (kT/\epsilon)\left|\ln T/T_1\right| \tag{29}$$

where $\epsilon = DK_2^2$ and $T_1 = D\epsilon^2/e^2$.

The lnT singularity in (29) comes about because a $\ln|\omega - E_m|$ type singularity appears in Γ_m. As discussed after (14), we must evaluate Γ at $\omega = \tilde{E}_m$. If we ignore the shift in the self energy and set $\omega = E_m$ we would have obtained a genuine logarithmic singularity. This is avoided by including the shift Δ_m given in (7) which leads to the cut-off T_1. For parameters appropriate for Si-MOSFET, it is found that the $|\ln T/T_1|$ factor can be of order 10. Thus even for moderate amounts of disorder ($k_F\ell \gtrsim 1$), Γ^{2D} can be larger than kT due to the enhancement by the $|\ln T/T_1|$ factor. Recently the inelastic scattering rate Γ in Si-MOSFET was extracted by the analysis of magnetoresistance data by BISHOP, DYNES and TSUI [12]. They find that Γ is roughly proportional to kT and scales with the diffusion constant as predicted by (29). Furthermore even the absolute magnitude is given correctly by (29) and that for most cases $\Gamma \gtrsim kT$.

Eqs. (28) and (29) are for strictly 2D films. In thin films of thickness t which is large compared with the screening length, the 3D screened interaction should be used. Furthermore, in the momentum integral in (25), it is the momentum transfer satisfying $Dq^2 < \max(E,kT)$ that is important. Thus there will be a cross-over from 2D to 3D behavior when the temperature or frequency exceeds Ω_c given by

$$t = \sqrt{D/\Omega_c}\ . \tag{30}$$

Furthermore, in the 2D regime, (28) and (29) must be multiplied by the factor $\pi/k_F t$ which will account for both the phase space restriction of the thin film geometry and the 3D screening.

We note that the presence of the logarithmic term in the decay rate (Eqs. (28) and (29)) together with the fact that the quasi-particle fraction Z contains logarithmic divergence (to be discussed later) indicate that Fermi liquid theory begins to break down at and below two dimensions.

3. Hartree Terms

So far we have discussed the exchange contribution to the self energy. In the limit of a δ function interaction, it is clear that the Hartree term with parallel spin and the exchange term are equal in magnitude and opposite in sign [13,14]. Diagrammatically the Hartree version of Fig. 2 is shown in Fig. 3.

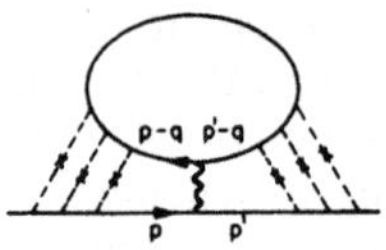

Figure 3

Physically the Hartree term is the interaction of a given eigenfunction with the nonuniform electron density in the ground state. Singularities appear because wavefunctions nearby in energy are correlated in space. Several differences from the exchange term should be noted. The Hartree term requires zero frequency transfer, so that only the static limit of the screened Coulomb interaction is involved. Furthermore the momentum transfer in the interaction is not dominated by small q because unlike the exchange term it is not the same as the momentum appearing in the diffusion pole. Thus for a static screened interaction v(q), the Hartree term is reduced relative to the static exchange term by a factor F

$$F = \frac{\int d\hat{\Omega}\; v(q=2k_F \sin\theta/2)}{\int d\hat{\Omega}\; v(0)} \tag{31}$$

which is the average of the interaction on the Fermi surface over the solid angle $\hat{\Omega}$.

In addition to the diagrams shown in Figs. 2 and 3 there are diagrams which involve singularities in the particle-particle channels. The contribution to these diagrams comes from large momentum transfer in the interaction ($q\ell > 1$) and should also be reduced to the factor F. Furthermore, the particle-particle channel singularity is cut-off by a relatively weak magnetic field, i.e. when

$$L_H \lesssim (D/\Omega_0)^{1/2} \tag{32}$$

where $L_H = (\hbar c/eH)^{1/2}$ is the Landau orbit size and Ω_0 is ω, kT or the inelastic scattering rate τ_{in}^{-1}, which ever is greater. These diagrams are discussed in detail by FUKUYAMA [4,15]. In the following we assume that these diagrams have been suppressed by a magnetic field or other time reversal symmetry breaking mechanisms [14]. Experimentally the existence of these terms can be verified by measuring the change of the tunneling density of states in the presence of a magnetic field when the condition given by (32) is satisfied.

4. Specific Heat and Tunneling Density of States

As discussed in Section II.2 the dynamic screening introduces a frequency dependent $\tilde{\Sigma}(\omega)$ which means that the quasi-particle fraction Z defined in (10) deviates from unity. Specific heat measurements are insensitive to Z and are controlled by the energy level spacing, i.e. N_Q as defined by (13). The specific heat is computed from the free energy which is given by the standard expression

$$\Delta F = \frac{1}{2} T \sum_{\omega q} \int_0^1 d\lambda \, \frac{v_0(q)\Pi(q,\omega)}{1-\lambda v_0(q)\Pi(q,\omega)}$$

where $v_0(q)$ is the bare Coulomb interaction and $\Pi(q,\omega) = -(dn/d\mu)Dq^2/(Dq^2-i\omega)$ is the polarizability. It is easy to show that the static calculation given in Section II.1 gives the correct answer for N_Q, if we replace $v(0)(dn/d\mu)$ by unity. Combined with the Hartree term the specific heat correction δC is predicted to be

$$\delta C/C = \frac{(1-2F)}{4\pi^2(dn/d\mu)D} \left[\frac{T}{2D}\right]^{1/2} \quad (3D) \,, \tag{33a}$$

$$= \frac{1}{\pi m D}(1-2F)\ln T \quad (2D) \,. \tag{33b}$$

The factor of 2 in 2F denotes spin degeneracy. On the other hand, the single particle density of states per spin

$$\delta N_T(E) = \pi^{-1} \int d\vec{r} \, \mathrm{Im} G(\vec{r},\vec{r},E) \tag{34}$$

contains also the Z factor modification. This is the density of states measured in tunneling experiments. The results are

$$\frac{\delta N_T(E)}{N_0} = \frac{(2-2F)}{4\pi^2(dn/d\mu)D} \left[\frac{E}{2D}\right]^{1/2} \quad (3D) \,. \tag{35}$$

In strictly 2D we have to use the 2D screening function. It turns out that the singularity is dominated by the inelastic processes that contribute to Z so that the leading singularity is given by

$$\frac{\delta N_T(E)}{N_0} = -\frac{1}{2\pi m D} \ln|E\tau| \ln\frac{|E|}{DK_2^2} \,. \tag{36}$$

For a thin film of thickness t greater than the screening length K_2^{-1}, we have 3D screening and just as for the decay rate, we have

$$\frac{\delta N_T(E)}{N_0} = (2-2F)\frac{\pi}{k_F t}\,\frac{1}{\pi m D}\ln|E\tau| \,. \tag{37}$$

As pointed out by IMRY and OVADYAHU [16], the thickness dependence in (37) has the satisfying feature that the corrections scale with the resistance per square of the film, rather than the resistivity of the material. The cross-over from logarithmic to square root singularity as the film thickness increases has been observed experimentally [16].

III. CONDUCTIVITY AND MAGNETORESISTANCE

The conductivity can be calculated using the Kubo formula to lowest order in the interaction. From the diagrams involving particle-hole diffusion alone, the following correction to the conductivity is obtained [3,4,13,14]

$$\delta\sigma_I = \frac{e^2}{\hbar}\frac{1}{4\pi^2}(2-2F)\ln(T\tau) \quad (2D) \tag{38a}$$

$$= \frac{e^2}{\hbar}\frac{1}{4\pi^2}\frac{1.3}{\sqrt{2}}(\frac{4}{3}-2F)\sqrt{\frac{T}{D}} \quad (3D)\ . \tag{38b}$$

The factor of 2 in front of F is for spin degeneracy. We should also point out that the dynamic (imaginary) part of $v(q,\omega)$ makes a contribution equal to the static part in 2D and 1/3 of the static part in 3D. This accounts for the factor 2 and 4/3 in Eqs. 38a and b respectively.

It is now known that even a weak magnetic field can suppress the localization effect [14]. The origin of that effect is that the particle-particle channel is sensitive to the magnetic flux. The interaction effects included in (38) are all based on the particle-hole diffusion channel, so that similar sensitivity to the magnetic field does not occur. There are also interaction effects due to the particle-particle channel. The sensitivity to magnetic field of these terms has been mentioned earlier (see Eq. (32)) and the implication for magnetoresistance has been discussed by FUKUYAMA [15,17] and ALTSHULER et al. [18]. As far as the particle-hole contribution in (38) is concerned, the only magnetic field effect is due to the splitting of the spin up and down bands (LEE and RAMAKRISHNAN [19], KAWABATA, [20]). The physical idea is most simply illustrated for the self-energy correction. As discussed in Section III.b, the singular correction is due to the correlation between the wavefunctions of the added electron with the wavefunctions of the occupied electrons that are nearby in energy. The exchange and the equal spin Hartree terms involve correlation with electrons with the same spin, and are unaffected by the spin splitting. This leaves the Hartree term between opposite spins and the spin splitting produces a gap $g\mu_B H$ between the lowest unoccupied spin up electron and the highest occupied spin down electron. The singularity of that term is therefore cut-off for $g\mu_B H$ greater than kT. In a magnetic field the correction to the conductivity can be written as a sum of two terms

$$\delta\sigma_I(H,T) = \delta\sigma_I'(T) + \delta\sigma_I''(H,T)\ . \tag{39}$$

The first term $\delta\sigma_I'$ is the field independent exchange and equal spin Hartree contribution and is the same as (38) except that the factors (2-2F) and (4/3-2F) are replaced by 2-F and 4/3-F respectively. The second term is the opposite spin Hartree contribution. Its field dependence is given by

$$\delta\sigma_I''(H,T) - \delta\sigma_I''(0,T) = -\frac{e^2}{\hbar}\frac{F}{4\pi^2}g_2(h) \quad (2D) \tag{40a}$$

$$= -\frac{e^2}{\hbar}\frac{F}{4\pi^2}\sqrt{\frac{T}{2D}}g_3(h) \quad (3D) \tag{40b}$$

where

$$g_2(h) = \int_0^\infty d\Omega \frac{d^2}{d\Omega^2}(\Omega N(\Omega)) \ln\left|1-\frac{h^2}{\Omega^2}\right| \tag{41a}$$

in 2D and

$$g_3(h) = \int_0^\infty d\Omega \frac{d^2}{d\Omega^2}(\Omega N(\Omega))(\sqrt{\Omega+h}+\sqrt{|\Omega-h|}-2\sqrt{\Omega}) \tag{41b}$$

in 3D and $h = g\mu_B H/kT$, $N(\omega) = (\exp(\omega)-1)^{-1}$. The zero field contribution $\delta\sigma_I''(0,T)$ is the usual one, and is given by (38) with (2−2F) and (4/3−2F) replaced by −F. The functions g_2 and g_3 can be computed numerically. They have the limiting behavior

$$g_2 = \begin{cases} \ln(h/1.3) & h \gg 1 \\ 0.084h^2 & h \ll 1 \end{cases} \tag{42a}$$

and

$$g_3 = \begin{cases} \sqrt{h}-1.3 & h \gg 1 \\ 0.053h^2 & h \ll 1 \end{cases} \quad . \tag{42b}$$

In (42), the constant 1.3 in the $h \gg 1$ limit is simply the $H = 0$ term subtracted in (40). It is worth noting that in the large field limit, $h \gg 1$, $\delta\sigma_I''(H,T)$ becomes temperature independent. The only temperature dependence due to interactions is due to $\delta\sigma_I'$ and the coefficient is proportional to 4/3−F which is always positive. This is in contrast to the $H = 0$ case when the $\sqrt{T}$ coefficient can change sign depending on the value of F which varies between zero and unity.

In 3D the low temperature $\sqrt{T}$ behavior of the conductivity has been observed in doped semiconductor such as Si:P [21], and in ferromagnetic metallic glasses [22]. A positive magnetoresistance proportional to $H^{1/2}$ has also been observed in Si:P [23] and in granular Al [24].

In 2D it is possible to separate the spin and orbital contribution to magnetoresistance because the spin part is isotropic while the orbital part requires a component of H normal to the film. Such a spin dependent magnetoresistance is recently observed in Si-MOSFET by BISHOP et al. [12].

IV. CONCLUSION

Interaction effects in disordered systems lead to surprisingly singular corrections to electronic density of states and to transport properties. Many of these effects have been observed experimentally. Our discussion has been limited to the weakly interacting and weakly disordered region. A proper understanding of the metal-insulator transition region clearly requires a treatment of interaction and localization effects on equal footing. Much work remains to be done in this direction.

REFERENCES

1. P. Nozieres and D. Pines, "Theory of Quantum Liquids", Benjamin, NY 1966.
2. O. Betbeder-Matibet and P. Nozieres, Annals of Physics *37*, 17 (1966).
3. B. L. Altshuler and A. G. Aronov, Solid State Comm. *30* 115 (1979); JETP *50*, 968 (1979).
4. H. Fukuyama, J. Phys. Soc. Jpn. *48*, 2169 (1980).
5. W. F. Brinkman and S. Engelsberg, Phys. Rev. Lett. *21*, 1187 (1968).
6. A. A. Abrikosov, L. Gorkov and I. Dzyaloshinskii, "Quantum Field Theoretical Methods in Statistical Physics", Pergamon 1965.
7. E. Abrahams, P. W. Anderson, D. C. Licciadello and T. V. Ramakrishnan, Phys. Rev. Lett. *42* 673 (1979).
8. D. Foerster, "Hydrodynamic Fluctuations, Broken Symmetry and Correlation Functions", Benjamin 1975.
9. E. Abrahams, P. W. Anderson, P. A. Lee and T. V. Ramakrishnan, Phys. Rev. B
10. P. F. Maldague, Phys. Rev. B *23*, 1719 (1981).
11. A. Schmid, Z. Phys. *271*, 251 (1974).
12. O. Bishop, R. C. Dynes and D. Tsui, unpublished.
13. B. L. Altshuler, A. G. Aronov and P. A. Lee, Phys. Rev. Lett. *44*, 1288 (1980).
14. B. L. Altshuler, D. Khmel'nitskii, A. I. Larkin and P. A. Lee, Phys. Rev. B*22*, 5142 (1980).
15. H. Fukuyama, these conference proceedings.
16. Y. Imry and Z. Ovadyahu, preprint.
17. H. Fukuyama, preprint.
18. B. L. Altshuler, A. G. Aronov, A. Z. Larkin and D. E. Khmel'nitskii, Zh. Eksp. Teor. Fiz. *81*, no. 8 (1981).
19. P. A. Lee and T. V. Ramakrishnan, preprint.
20. A. Kawabata, preprint.
21. T. Rosenbaum, K. Andres, G. A. Thomas, and P. A. Lee, Phys. Rev. Lett. *46*, 568 (1981).
22. D. Rapp, S. M. Bhagat and H. Gudmundsson, preprint; O. Rapp, S. M. Bhajat and Ch. Johannesson, Sol. St. Comm. *21*, 83 (1977).
23. T. Rosenbaum et al., Phys. Rev. Lett. *47*, 1758 (1981).
24. T. Chui, P. Lindenfeld, W. L. McLean and K. Mui, Phys. Rev. Lett. *47*, 1617 (1981).

Interaction Effects in Two-Dimensional Disordered Systems

H. Fukuyama

The Institute for Solid State Physics, The University of Tokyo
Roppongi, Minato-ku, Tokyo 106, Japan

Interaction effects in the weakly localized regime in two-dimension are reviewed. Especially the magneto-resistance due to orbital motions combined with interactions is examined in detail and the theoretical results are shown to explain very nicely experiments in MOS and very metallic Cu films. In the case of MOS it is seen that the intervalley impurity scattering plays essential roles. It is also pointed out that experiments in some metallic films can be qualitatively understood if strong spin-orbit interactions are taken into account.

1. Introduction

Recent scaling theory by ABRAHAMS et al. [1] and by ANDERSON et al. [2] has made for the first time microscopic investigations possible on the localization. Especially the weakly localized regime in two dimensions has been playing important roles to check new theories [3]. The characteristic feature in this regime, where precursor effects to the complete localization are seen, is the logarithmic dependences of conductivity on temperature, magnetic field and electric field. The prefactors of such logarithmic dependences reflect subtle natures of the system and then detailed analysis of this regime affords us information which is not expected otherwise. Especially the mutual interaction treated in its linear order is seen to affect these prefactors [4-7]. Such interaction effects have been characterized [5,6] by four different effective coupling constants, g_i (i=1,2,3 and 4), depending on the Hartree or the Fock type processes modified by the diffusion process either of particle-particle or of particle-hole type. These coupling constants affect the temperature and the magnetic field dependences in a different way, and then the critical check of theoretical predictions has been possible.

In this paper a review is given on the present status of theoretical investigations on interaction effects in two dimensions in view of experiments on Si-MOS [8,9] and various metallic films [9]. (We take unit of $\hbar=k_B=1$.) The interaction effects including three-dimensional systems will also be discussed by Lee in these proceedings.

2. Single-Valley System

The simplest theoretical model for microscopic investigations is

$$\mathcal{H} = \sum_i \left[\frac{p_i^2}{2m} + \sum_{R_\ell} u\delta(r_i - R_\ell)\right] + \frac{1}{2}\sum_{i\neq j} v(r_i - r_j)$$

$$= \mathcal{H}_0 + V \qquad , \qquad (2.1)$$

where the randomness is characterized as a short-range impurity potential with strength, u, and v(r) is the effective mutual interaction. In this model the lifetime τ of the Bloch state due to elastic scattering u is given by

$$\tau = (2\pi N(0) n_i u^2)^{-1} \qquad , \qquad (2.2)$$

where n_i is the average impurity density and $N(0)=m/2\pi$ is the density of states per spin.

In the localization theory, which does not consider the interaction effects explicitly but takes these into account by the phenomenological parameter of the inelastic scattering time τ_ε [2], the quantum corrections to the conductivity, σ_L', is given as follows [2],

$$\sigma_L' = -\frac{e^2}{2\pi^2} \ln \tau_\varepsilon/\tau \qquad , \qquad (2.3)$$

or

$$\sigma_L'/\sigma_0 = -\frac{1}{2\pi\varepsilon_F\tau} \ln \tau_\varepsilon/\tau \qquad , \qquad (2.4)$$

$$\sigma_0 = e^2\varepsilon_F\tau/\pi \qquad , \qquad (2.5)$$

where ε_F is the Fermi energy and σ_0 is the classical conductivity obtained from ordinary Boltzmann transport equation. This contribution, σ_L', is due to the maximally crossed diagram, or, the particle-particle diffusion propagators given below by eq.(2.8b) [10,11].

The interaction theories, on the other hand, result in [5,6]

$$\sigma_I' = -\frac{e^2}{2\pi^2} g \ln \frac{1}{\tau T} \qquad , \qquad (2.6)$$

$$g = g_1 + g_2 - 2(g_3 + g_4) \qquad , \qquad (2.7)$$

where $g_i = v(q,\omega_\ell)N(0)$ (i=1,2,3 and 4), $v(q,\omega_\ell)$ being the Fourier transform of the effective mutual interaction with thermal frequency $\omega_\ell = 2\pi\ell T$, are dimensionless coupling constants characterizing four distinct processes defined in Figs.1(a)-(d) for the self-energy corrections. In these figures dotted lines with crosses mean to take average over impurity configurations whereas broken and double-broken lines represent particle-hole, $D(q,\omega_\ell)$, and particle-particle, $D_\varepsilon(q,\omega_\ell)$, diffusion propagators, respectively, and are defined in Figs.2(a) and (b). These propagators in case of $\varepsilon_n(\varepsilon_n+\omega_\ell)<0$, where $\varepsilon_n=(2n+1)\pi T$, are given by

$$D(q,\omega_\ell) = \frac{1}{2\pi N(0)\tau^2}\frac{1}{Dq^2+|\omega_\ell|} , \tag{2.8a}$$

$$D_\varepsilon(q,\omega_\ell) = \frac{1}{2\pi N(0)\tau^2}\frac{1}{Dq^2+|\omega_\ell|+\frac{1}{\tau_\varepsilon}} . \tag{2.8b}$$

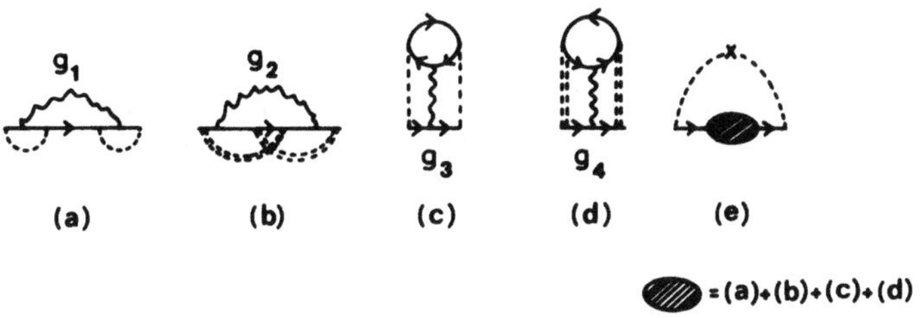

Fig. 1 Definitions of the effective coupling constants, g_i(i=1,2,3 and 4) in terms of the self-energy corrections. Here broken and double broken lines are particle-hole and particle-particle diffusion propagators defined in Fig.2(a) and (b), respectively.

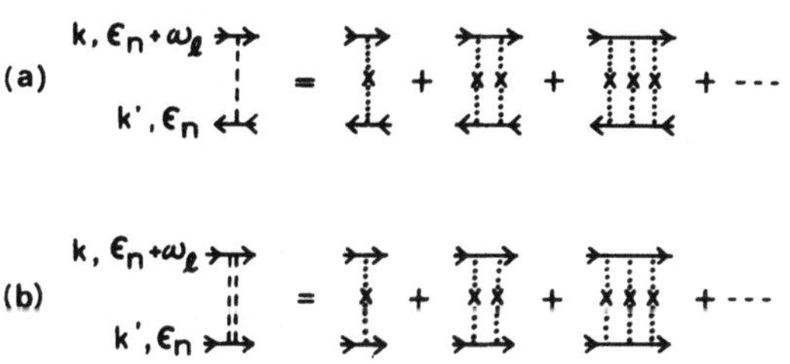

Fig. 2 The particle-hole diffusion propagator (a) and the particle-particle diffusion propagator (b).

Here $D=\varepsilon_F\tau/m$ is the diffusion constant. We see that the energy transfer ω_ℓ through interactions are small, $|\omega_\ell|\tau<<1$, in all of the processes. The momentum transfer q, however, depends on process: in the g_1-process $|q|$ is small, i.e. $|q|<q_c\equiv 1/\sqrt{D\tau}$, whereas $|q|$ can be large in other processes. It should be noted that processes with particle-particle diffusion propagators, i.e. g_2 and g_4, affect the conductivity in the same order of magnitude as particle-hole diffusion processes. In the calculation of ALTSHULER et al. [4], only g_1 and g_3 processes have been taken into account. The existence of contributions from g_2 and g_4 processes has also been noted by ALTSHULER et al. [7] but they did not perform explicit calculations of the quantum corrections to the conductivity due to these interactions.

If $\tau_\varepsilon\propto T^{-p}$ (p>0) [2], eqs.(2.3) and (2.6) indicate that the prefactor of the logarithmic temperature dependence is given by

$$\sigma' \equiv \sigma_L' + \sigma_I' = \frac{e^2}{2\pi^2}\alpha_T \ell n T\tau , \tag{2.9}$$

$$\alpha_T = p + g \quad , \tag{2.10}$$

where g is given by eq.(2.7). If dynamically screened Coulomb interaction is assumed [4], i.e.

$$v(q,\omega_\ell) = \frac{2\pi e^2}{\varepsilon_0 q} \frac{Dq^2+|\omega_\ell|}{D\kappa_0 q+|\omega_\ell|} \quad , \tag{2.11}$$

where ε_0 is the dielectric constant and $\kappa_0=4N(0)e^2/\varepsilon_0$ is the inverse screening radius, g_i's in eq.(2.7) are as follows:

$$g_1 = 1 \quad , \tag{2.12a}$$

$$g_2 = g_3 = g_4 = F/2 \quad . \tag{2.12b}$$

Here quantity F is introduced by ALTSHULER et al. [4] and is defined as $F=2\langle v(k-k',\omega_\ell)N(0)\rangle \simeq 2\langle v(k-k',0)N(0)\rangle$ averaged over two wavevectors, k and k', on the Fermi surface. It is easily seen that $F\to 1$ as $2k_F/\kappa_0 \to 0$ and $F\to 0$ as $2k_F/\kappa_0 \to \infty$.

As regards the magnetic field dependences we note that the particle-particle diffusion propagator, $D_\varepsilon(q,\omega_\ell)$, is sensitive to the field, H, applied perpendicularly to the systems, since q in eq.(2.8b) is affected by the vector potential, A, as $q\to q-2eA/c$ [12]. Consequently contributions in the localization theory, σ_L', [7,13] and those from g_2- and g_4- processes [14] yield magnetoresistance (MR) due to orbital motions [15]. Besides this orbital effect it was recently noted [16,17] that the spin Zeeman splitting also affects σ_I'. (Similar effects of spin Zeeman splitting on the particle-particle diffusion propagators had been noted in the discussions of MR in the presence of spin-orbit scattering but without interactions [18].) However the characteristic strength of the field to observe these orbital and spin effects, H_0 and H_s, respectively, are very different; for orbital effects $H_0 \sim /(c/De\tau_\varepsilon) = (mc/e)\cdot(1/\varepsilon_F \tau\tau_\varepsilon)$ and for spin effects $H_s \sim T/\mu_B$, μ_B being the Bohr magneton. Since $H_0 \ll H_S$ if $\varepsilon_F\tau \gg 1$, the orbital effects dominate over the spin effects in the field region $H\sim H_0$. In the following discussion, except in Sec.4, we restrict ourselves to such field region and then we ignore spin effects. In this case the changes of conductivity due to magnetic fields are given by

$$\Delta\sigma_L' \equiv \sigma_L'(H) - \sigma_L'(0) = \frac{e^2}{2\pi^2}\Phi(h) \quad , \tag{2.13}$$

$$\Delta\sigma_I' \equiv \sigma_I'(H) - \sigma_I'(0) = \frac{e^2}{2\pi^2}(g_2-2g_4)\,\phi(h,\gamma) \quad , \tag{2.14}$$

where $\Phi(h)$[7,13] and $\phi(h,\gamma)$[14] are defined as follows, respectively:

$$\Phi(h) = \psi(\frac{1}{2} + \frac{1}{a\tau_\varepsilon}) - \psi(\frac{1}{2} + \frac{1}{a\tau}) + \ln \tau_\varepsilon/\tau$$

$$\sim \psi(\frac{1}{2} + \frac{1}{h}) + \ln h \quad , \tag{2.15}$$

$$\phi(h,\gamma) = - \sum_{\ell=1}^{\infty} [5\{\frac{\ell}{(\gamma h)^2}\psi''(\frac{1}{2}+\frac{1}{h}+\frac{\ell}{\gamma h}) + \frac{\ell}{(\ell+\gamma)^2}\}$$

$$+ 6\{\frac{\ell(\ell+\gamma)}{2(\gamma h)^3}\psi'''(\frac{1}{2}+\frac{1}{h}+\frac{\ell}{\gamma h}) - \frac{\ell}{(\ell+\gamma)^2}\}] \quad . \tag{2.16}$$

Here $a=4DeH/c$, $h=a\tau_\varepsilon$, $\gamma=(2\pi\tau_\varepsilon T)^{-1}$ and $\psi^{(n)}(z)$ is the poly-gamma function. In the field region where $h\gg1$ and $h\gamma\gg1$, both $\Phi(h)$ and $\phi(h,\gamma)$ depend on h logarithmically, i.e.

$$\Phi(h) \sim \phi(h,\gamma) \sim \ln h \quad . \tag{2.17}$$

If eq.(2.17) applies, we obtain

$$\Delta\sigma' \equiv \Delta\sigma_L' + \Delta\sigma_I' = \frac{e^2}{2\pi^2}\alpha_H \ln h \quad , \tag{2.18}$$

$$\alpha_H = 1 + g_2 - 2g_4 \quad . \tag{2.19}$$

From eqs.(2.10) and (2.19) we see that the prefactors of logarithmic dependences on temperatures and magnetic fields are as follows when the dynamically screened Coulomb interactions are assumed between electrons,

$$\alpha_T = p + 1 - 3F/2 \quad , \tag{2.20}$$

$$\alpha_H = 1 - F/2 \quad . \tag{2.21}$$

It should be noted that these results have been derived for a <u>single</u> kind of electron in a disordered system.

The typical experimental example in the present context appears to be very metallic Cu films by Van den DRIES et al. [19], where the sheet resistance, $R_\square$, is $10\Omega \lesssim R_\square \lesssim 100\Omega$ and $\alpha_T \sim 2$ and $\alpha_H \sim 0.75\pm0.5$ are observed. The theoretical value of F, on the other hand, is quoted in ref.[19] to be $F\sim0.6$. Thus eq.(2.21) explains MR, whereas eq.(2.20) is in agreement with the experiment if $p\sim2$.

3. Many-Valley System

The typical example of two-dimensional systems will be Si-MOS, where there exists the valley degeneracy, n_v, depending on the orientation of the silicon surface; $n_v=2$ in (1,0,0) surface whereas $n_v=6$ in ideal (1,1,1) surface, though the valley degeneracy in the latter case appear to be not yet so definite experimentally [20].

If there exist n_v independent and equivalent valleys with the isotropic mass, we will have α_T and α_H defined by eqs.(2.9) and (2.18), respectively, as follows in the case of dynamically screened Coulomb interactions,

$$\alpha_T = n_v\,[p + (1 + F(\tfrac{1}{2} - 2n_v))/n_v] \quad , \tag{3.1}$$

$$\alpha_H = n_v\,[1 + F(\tfrac{1}{2} - n_v)/n_v] \quad . \tag{3.2}$$

In deriving eqs.(3.1) and (3.2), we noted that the inverse screening radius in the presence of n_v valleys is larger by n_v than that in a single valley.

However in the actual experiments in Si-MOS such independence of valleys will not be justified. From the detailed analysis of the structure of the conductivity associated with the minigap, which is sensitive to the intervalley scattering, ANDO [21] concluded that the intervalley scattering is playing an important role, i.e. some fraction of the mobility is determined by such scattering.

In this section we will see that α_T and α_H are essentially affected by such intervalley impurity scatterings [6,22-24]. The physical parameter to determine α_T and α_H is seen to be τ_ε/τ', where τ' is the lifetime due to the intervalley scattering. Thus even though intervalley scattering is weak in the sense of $\tau' \gg \tau_0$, where τ_0 is lifetime due to intravalley scattering, α_T and α_H are modified from those in independent valleys as far as $\tau_\varepsilon \gg \tau'$. Namely the competition between relatively weak scattering mechanism determining τ_ε and τ' is clearly reflected in the prefactors of the logarithmic dependences. In the following we will evaluate the logarithmic prefactors in the limit of $\tau'/\tau_\varepsilon \to 0$, i.e. strong intervalley scattering compared to the inelastic scattering

As a typical example we consider the case with two equivallent valleys, i.e. $n_v=2$ as in Si-(1,0,0) surface. The model Hamiltonian is, instead of $\mathcal{H}_0$ in eq.(2.1),

$$\mathcal{H}_0 = \sum_k \frac{k^2}{2m} C^+_{1k}C_{1k} + \sum_k \frac{k^2}{2m} C^+_{2k}C_{2k}$$
$$+ \sum_k \sum_q u(q) \left(C_{1k+q}C_{1k} + C_{2k+q}C_{2k}\right)$$
$$+ \sum_k \sum_q \left(v(q)\, C_{1k+q}C_{2k} + \text{h.c.}\right) \quad . \tag{3.3}$$

Here 1 or 2 refers to the valley indices and u(q) and v(q) are scattering potentials in and between valleys, respectively. We assume that the random average of $\langle u(q)v(q')\rangle$ is vanishing due to the fact that v(q) is a rapidly oscillating function with the wave vector connecting two valley centers. The life times due to intravalley and intervalley scatteirng, τ_0 and τ', are defined as

$$\langle u(q)u(q')\rangle = \delta_{q,q'} \frac{1}{2\pi N(0)\tau_0} \quad , \tag{3.4a}$$

$$\langle v(q)v(q')\rangle = \delta_{q,q'} \frac{1}{2\pi N(0)\tau'} \quad , \tag{3.4b}$$

and total elastic scattering time τ is given by

$$\frac{1}{\tau} = \frac{1}{\tau_0} + \frac{1}{\tau'} \quad . \tag{3.5}$$

In the presence of the intervalley impurity scattering, particle-particle and particle-hole diffusion propagators have following characteristic features depending on the valley indices. One can think of three different types of particle-particle diffusion propagators as shown in Figs.3(a)-(c), where a and b($\neq$a) are valley indices. As is explicitly evaluated in ref.[24], the process in Fig.3(c) is not singular as $q\to 0$, $\omega_\ell \to 0$, whereas those in Fig.3(a) and (b), defined as $\Lambda_0(q,\omega_\ell)$ and $\Lambda_1(q,\omega_\ell)$, respectively, are given as

$$\Lambda_0(q,\omega_\ell) = \Lambda_1(q,\omega_\ell) = \frac{1}{2} \times \frac{1}{2\pi N(0)\tau^2} \frac{1}{Dq^2+|\omega_\ell|} \quad . \tag{3.6}$$

Note that the prefactor is reduced by 2 compared to eq.(2.8b). On the other hand there exist three different types of particle-hole diffusion propagators

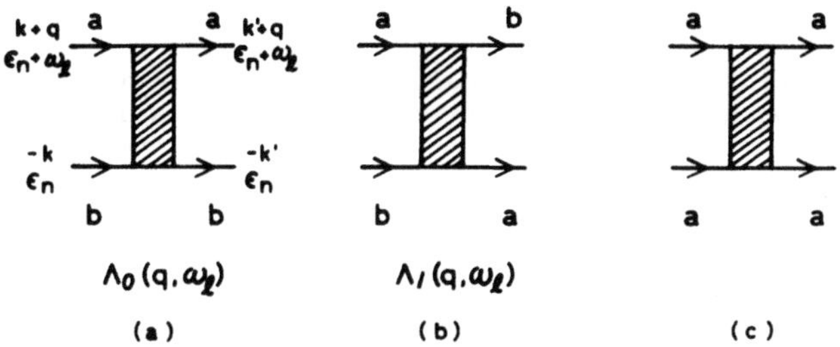

Fig. 3 Various possible particle-particle diffusion processes in two-valley systems where a and b($\neq$a) are valley indices. Only $\Lambda_0(q,\omega_\ell)$ and $\Lambda_1(q,\omega_\ell)$ have singularities as $q\to0$, $\omega_\ell\to0$.

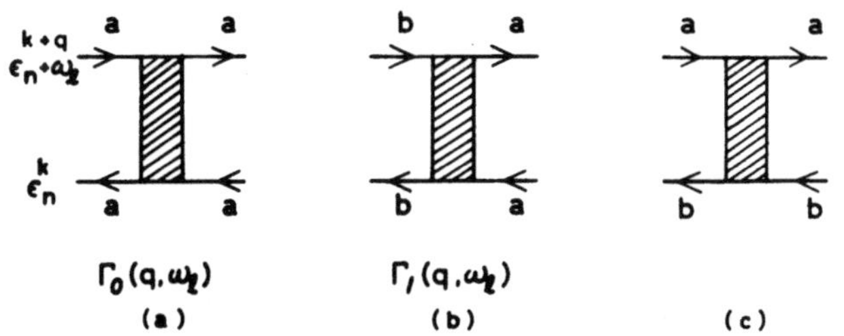

Fig. 4 Various possible particle-hole diffusion processes for two-valley systems. Only $\Gamma_0(q,\omega_\ell)$ and $\Gamma_1(q,\omega_\ell)$ have singularities $q\to0$, $\omega_\ell\to0$.

as shown in Fig.4(a)-(c), among which Fig.4(c) is not singular, whereas Figs. 4(a) and (b) are singular and are given as follows,

$$\Gamma_0(q,\omega_\ell) = \Gamma_1(q,\omega_\ell) = \frac{1}{2}\times\frac{1}{2\pi N(0)\tau^2}\;\frac{1}{Dq^2+|\omega_\ell|} \quad . \tag{3.7}$$

By use of eqs.(3.6) and (3.7) the quantum corrections to the conductivity are evaluated. The contribution in the localization theory, σ_L', is determined by $\Lambda_1(q,\omega_\ell)$ and the result is

$$\sigma_L' = -\frac{e^2}{2\pi^2}\,\ell n\;\tau_\varepsilon/\tau \quad . \tag{3.8}$$

This contribution is in contrast to that in independent valleys, where one expects

$$\sigma_L' = -\frac{e^2}{2\pi^2}n_v\;\ell n\;\tau_\varepsilon/\tau \quad . \tag{3.9a}$$

Comparison of eq.(3.8) with eq.(3.9a) implies that, if one introduces the parameter, α, characterizing the effects of intervalley scattering by writing

$$\sigma_L' = -\frac{e^2}{2\pi^2}n_v\alpha\;\ell n\;\tau_\varepsilon/\tau \quad , \tag{3.9b}$$

one obtains $n_v\alpha \to 1$ in the limit of $\tau'/\tau_\varepsilon \to 0$. This result of $n_v\alpha=1$ in this limit has been explicitly obtained not only for the case of $n_v=2$ but also for $n_v=6$ and appears to be quite general [23]. The physical implication of $n_v\alpha=1$ is that electrons in n_v valleys experience inelastic scattering as a single kind of carrier due to the rearrangement of electronic states by intervalley impurity scattering.

The contributions due to interactions in this limit of $\tau'/\tau_\varepsilon \to 0$ have been evaluated in ref.[24] and the result for $n_v=2$ is as follows:

$$\sigma_I' = -\frac{e^2}{2\pi^2} g_{II} \ln \frac{1}{\tau T} , \tag{3.10}$$

$$g_{II} = 2g_1 + g_2 - 2(g_3+g_4) . \tag{3.11}$$

In the case of the dynamically screened Coulomb interaction, g_i, in the present case of $n_v=2$ are reduced by 2 compared to those in the single valley system, eqs.(2.12a) and (2.12b), i.e.

$$g_1 = 1/2 , \tag{3.12a}$$

$$g_2 = g_3 = g_4 = \frac{1}{8\pi} \int_0^{2\pi} d\theta \frac{\kappa}{\kappa + 2k_F \sin\frac{\theta}{2}} \equiv \frac{F}{4} , \tag{3.12b}$$

where $\kappa = 8\pi e^2 N(0)/\varepsilon_0$. For Si-(1,0,0) surface we have

$$\kappa = 1.87 \times 10^7 \text{ cm}^{-1} , \tag{3.13}$$

$$k_F = 1.77 \times 10^6 \sqrt{\nu} \text{ cm}^{-1} , \tag{3.14}$$

where ν is the electron density, N_s, in unit of 10^{12}cm^{-2}. The N_s dependence of F is shown in Fig.5. As is seen, $F \simeq 1$ in the ordinary experimental condition.

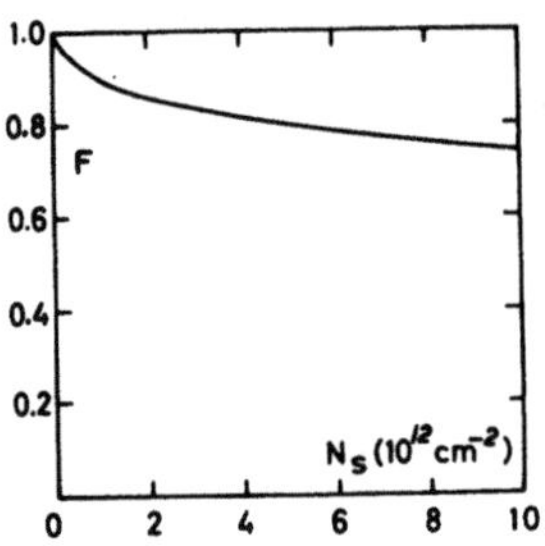

Fig. 5 Electron density (N_s) dependence of the effective interaction, F, defined by eq.(3.12b) for Si-(1,0,0) surface.

From eqs.(3.8) and(3.10) we found that the prefactor, α_T, of the logarithmic temperature dependence in the case of $n_v=2$ with strong intervalley scattering, $\tau' \ll \tau_\varepsilon$, is given as

$$\alpha_T = p + g_{II}$$
$$= p + 1 - 3F/4 , \tag{3.15}$$

in contrast to eq.(3.1) for independent valleys. On the other hand the prefactor, α_H, of the logarithmic field dependence is

$$\alpha_H = 1 + g_2 - 2g_4 = 1 - F/4 \quad . \tag{3.16}$$

Experimentally KAWAGUCHI and KAWAJI [25-27] determined the N_s-dependence of α_H for Si-(1,0,0). The result is shown in Fig.5 [27]. (Note that α defined in ref.[27] is related to our α_H by $\alpha_H=n_v\alpha$, i.e. $\alpha=\alpha_H/2$.) Since τ_ε will be increasing in proportion to ε_F and then to N_s whereas τ' will be weakly dependent on N_s, the intervalley impurity scattering will be relatively important at higher density. In such high density regions the experimental value of α_H is definitely smaller than 1 and indicates the existence of the interaction effect, i.e. the second term of r.h.s. of eq.(3.16). Since $F\sim 1$, the theoretical result is also in quantitative agreement with experiments.

The increase of α_H as N_s is decreased is qualitatively understood as the relative weakening of the intervalley impurity scattering: in the limit of independent valleys we expect, from eq.(3.1), $\alpha_H=2-F/2\sim 1.5$.

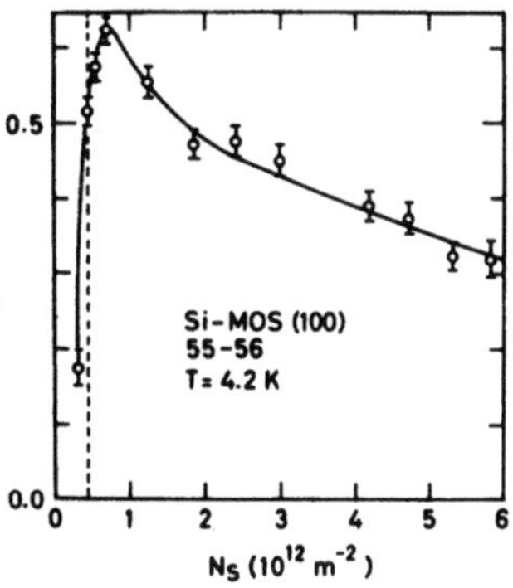

Fig. 6 Experimental result by Kawaguchi and Kawaji of electron density (N_s) dependence of α_H deduced from the magnetoresistance [27].

An interesting fact in the experiment of Fig.6 [27] is the sharp drop of α_H as N_s is decreased through $N_s\sim 5\times 10^{11}\,cm^{-2}$. This fact is consistent with the removal of the valley degeneracy, i.e. $n_v=2\to 1$, since

$$\alpha_H = 1 - F/2 \sim 0.5$$

when $n_v=1$ as seen from eq.(3.2). Although the crucial parameter, $\varepsilon_F\tau$, at this density of this sample is $\varepsilon_F\tau\sim 0.5$ and then $\varepsilon_F\tau \gg 1$ is not satisfied, we think it possible that such drastic effect as the removal of the valley degeneracy will be reflected in α_H. (The correct expansion parameter in this problem appears to be $\lambda=(2\pi\varepsilon_F\tau)^{-1}$ rather than $(\varepsilon_F\tau)^{-1}$ by use of the results by VOLLHARDT and WÖLFLE [28], and if this is the case, we have $\lambda\sim 0.3$ and then our theory will be applicable. Actually similar magnetic field dependences are seen at N_s below and above this critical density [27].)

This possibility of the removal of the valley degeneracy has been theoretically predicted by BLOSS, SHAM and VINTER [29] some time ago. The critical density observed in α_H is in agreement with their prediction. It is very interesting and important to realize that α_H is sensitive to the valley degeneracy. Further experimental investigations are admittedly necessary to check this interesting possibility.

4. Effects of Spin-Orbit Scattering

There now exist experiments on various metallic films which can be considered as two-dimensional systems since thickness is less than the diffusion length $\sqrt{D\tau_\varepsilon}$ [9]. In contrast to the negative magnetoresistance observed in Si-MOS and pure Cu in perpendicular fields, both signs of MR have been observed. Although the experimental possibilities have not fully been explored, the following are the remarkable features to be noted. In the case of Pd [30] and Au [31], for example, MR is positive even though the resistivity increases logarithmically as the temperature is lowered, i.e. $\alpha_T>0$ but $\alpha_H<0$. In such a case of $\alpha_H<0$, the anisotropy is very much weaker than in the cases of Si-MOS and very metallic Cu films and the values of α_T is smaller.

Such features are qualitatively understood as follows in terms of the strong spin-orbit scattering together with Coulomb interactions [33]. The physical parameter characterizing the relative importance of the spin-orbit scattering is $\tau_{s0}/\tau_\varepsilon$, where τ_{s0}^{-1} is the scattering rate due to the spin-orbit scattering. In order to see the essential aspect of the strong spin-orbit scattering we will examine the case in the limit of $\tau_{s0}/\tau_\varepsilon\to 0$ [33]. In this case diffusion processes are dependent on spin indices and among various propagators shown in Figs.7(a) and (b) only $\Gamma_{++,--}$, $\Gamma_{-+,+-}$, $\Lambda_{++,++}$ and $\Lambda_{-+,-+}$ are of diffusion type, i.e. singular as $q\to 0$, $\omega_\ell\to 0$, and they are given as follows,

$$\Gamma_{++,--} = \Lambda_{++,++} = \Lambda_{-+,-+} = \frac{1}{4\pi N(0)\tau^2}\,\frac{1}{Dq^2+|\omega_\ell|}\,, \tag{4.1}$$

$$\Gamma_{-+,+-} = -\frac{1}{4\pi N(0)\tau^2}\,\frac{1}{Dq^2+|\omega_\ell|} \tag{4.2}$$

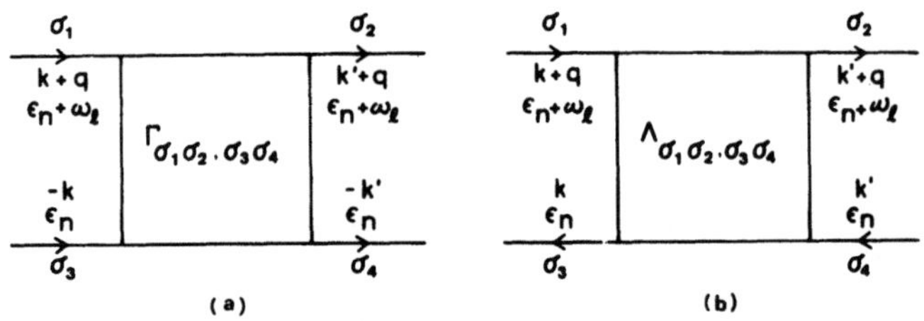

Fig. 7 Definition of particle-particle ($\Gamma_{\sigma_1\sigma_2,\sigma_3\sigma_4}$) and particle-hole ($\Lambda_{\sigma_1\sigma_2,\sigma_3\sigma_4}$) diffusion propagators in the presence of strong spin-orbit interactions.

It is to be noted that the absolute magnitude of these propagators are factor 2 smaller than those in eqs.(2.8a) and (2.8b) and, especially, that $\Gamma_{-+,+-}$ is negative.

As in the case of intervalley impurity scattering discussed in the previous section, where the valley indices played a role of spin indices here, processes yielding singular contributions to conductivity as $T\to 0$ have particular combinations of spin indices through scattering. The quantum correction in the absence of the interaction is determined by $\Gamma_{+-,-+}$ and $\Gamma_{-+,+-}$ and is given by [13]

$$\sigma_H' = \frac{e^2}{4\pi^2} \ln \tau_\varepsilon/\tau \quad . \tag{4.3}$$

The quantum corrections due to interactions are evaluated as follows [33]:

$$\sigma_I' = -\frac{e^2}{2\pi^2} g_{s0} \ln \frac{1}{4\pi\tau T} \quad , \tag{4.4}$$

$$g_{s0} = g_1 - \frac{g_2+g_3+g_4}{2} \quad , \tag{4.5}$$

where $2\pi\tau_\varepsilon T>1$ is assumed. Thus α_T in this case of $\tau_{s0}/\tau_\varepsilon \to 0$ is given by

$$\alpha_T = -p + g_{s0} \quad , \tag{4.6a}$$

$$= -p + 1 - 3F/4 \quad , \tag{4.6b}$$

provided $\tau_\varepsilon \propto T^{-p}$ and in eq.(4.6b) we assumed the dynamically screened Coulomb interactions. If τ_ε is weakly dependent on temperature due to, e.g., the possible existence of small amount of paramagnetic impurities, we will have $p\sim 0$ and then

$$\alpha_T \sim 1 - 3F/4 \quad . \tag{4.6c}$$

On the other hand α_H is given by

$$\alpha_H \sim -\frac{1}{2} - \frac{g_2+g_4}{2} \quad , \tag{4.7a}$$

$$= -\frac{1+F}{2} \quad . \tag{4.7b}$$

One interesting feature of MR is that it saturates in high fields to the same value irrespective of the direction of the field. This saturation value is given by

$$\Delta\sigma' = \sigma'(\infty) - \sigma'$$

$$= -\frac{e^2}{4\pi^2}\left[\ln\frac{\tau_\varepsilon}{\tau} + (g_2+g_4)\ln\frac{1}{4\pi\tau T}\right] \quad , \tag{4.8a}$$

$$= -\frac{e^2}{4\pi^2}\left[\ln\frac{\tau_\varepsilon}{\tau} + F\ln\frac{1}{4\pi\tau T}\right] \quad . \tag{4.8b}$$

This is in contrast to the case without spin-orbit scattering discussed in Sec.2, where one expects, including the spin Zeeman effects [16,17],

$$\Delta\sigma' = \sigma'(\infty) - \sigma'$$

$$= \frac{e^2}{2\pi^2}\left[\ln\frac{\tau_\varepsilon}{\tau} + (g_2 - g_3 - 2g_4)\ln\frac{1}{\tau T}\right] \tag{4.9a}$$

in perpendicular fields $H(>H_s>H_0)$ and

$$\Delta\sigma' = -\frac{e^2}{2\pi^2}(g_3+g_4)\,\ell n\frac{1}{\tau T} \qquad (4.9b)$$

in parallel fields $H(>H_s)$ [17].

The absence of the anisotropy at high fields in the presence of the strong spin-orbit scattering is due to the fact that all of the processes sensitive to the field result from the particle-particle diffusion propagators between antiparallel spins.

From eqs.(4.6c) and (4.7b) we see it possible to have $\alpha_T>0$ and $\alpha_H<0$ when τ_ε is weakly dependent on temperature. This appears to be the case in Pd [30, 38] and Au [31]. These are results in the case of strong spin-orbit scattering $\tau_{s0}/\tau_\varepsilon\to 0$.

Combining the present results and the results in the absence of spin-orbit scattering we schematically show the anisotropy of the field dependences of MR in Figs.8(a)-(c). The case of intermediate region corresponding to Fig.8(b) is Cu film with relatively large sheet resistance, as shown by KOMORI et al. [34], which has been analyzed in detail in [18], and Mg-Au alloys by BERGMANN [35].

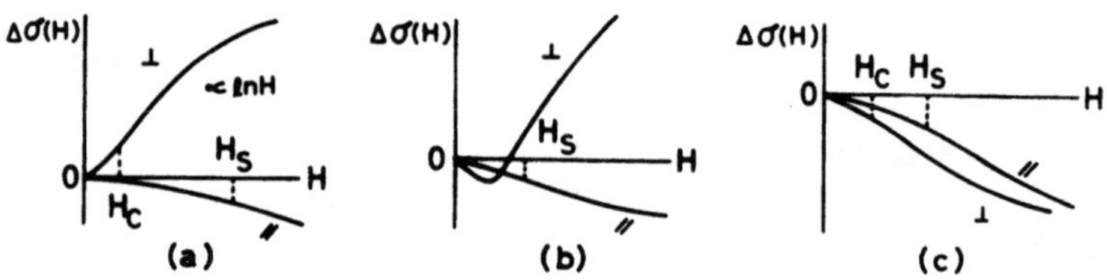

Fig. 8 Schematic representations of the anisotropies of magnetoresistance in the absence of (a), with intermediate (b) and strong (c) spin-orbit scattering.

5. Summaries and Discussion

Interaction effects in the weakly localized regime in two dimensions have been discussed. It is shown that four different coupling constants, g_i (i=1,2, 3 and 4), characterize all possible processes distinct from each other; the Hartree and the Fock processes with quantum corrections due to the particle-particle or the particle-hole diffusion propagators. Magnetoresistance due to both orbital and spin effects caused by these interactions is investigated.

We have seen that the theoretical results give satisfactory explanations for experiments on Si-MOS and various metallic films.

Important and interesting aspects of the interaction effects, which have not been discussed in this paper, will be the interplay between localization and various types of phase transition. Among these, the superconducting phase transition and the Kondo effect will be discussed by MAEKAWA and OHKAWA in this volume.

Throughout this paper we treated the quantum corrections to the conductivity due to interactions and localization on a different footing by introducing the phenomenological parameter, τ_ε, of the inelastic scattering lifetime for the localization theory. This can not be the complete story, since τ_ε itself is also to be determined once the interaction processes are given. Recently ABRAHAMS et al. [36] and LEE in this volume discussed the quasi-particle lifetime in the disordered systems in this context by examining the self-energy function of electrons. This lifetime is now seen [37] to be identified with τ_ε in the conductivity, σ_L', defined through the particle-particle diffusion propagators.

Acknowledgements

The author thanks Professor Kei Yosida for useful discussions in various stages of the work. He also appreciates informative discussions on experiments with Professors S. Kawaji and S. Kobayashi. Discussions with colleagues at Aspen, especially Drs. E. Abrahams, P.W. Anderson and P.A. Lee are also acknowledged.

References

1. E. Abrahams, P.W. Anderson, D.C. Licciardello and T.V. Ramakrishnan, Phys. Rev. Lett. 42, 673 (1979).
2. P.W. Anderson, E. Abrahams and T.V. Ramakrishnan, Phys. Rev. Lett. 43, 1718 (1979).
3. For a theoretical review, H. Fukuyama, Proc. 4th Intern. Conf. on Electronic Properties of Two-Dimensional Systems (New London, 1981) to appear in Surface Sci. 114 (1982).
4. B.L. Altshuler, A.G. Aronov and P.A. Lee, Phys. Rev. Lett. 44, 1288 (1980).
5. H. Fukuyama, J. Phys. Soc. Jpn. 48, 2169 (1980).
6. H. Fukuyama, Proc. Intern. Conf. Semiconductors (Kyoto, 1980), J. Phys. Soc. Jpn. 49, Suppl. A, 987 (1980).
7. B.L. Altshuler, D.E. Khemelnitzkii, A.I. Larkin and P.A. Lee, Phys. Rev. B22, 5142 (1980).
8. For a experimental review on MOS; M. Pepper, Proc. Intern. Conf. on Disordered Systems and Localizations (Rome, 1981).
9. For an experimental review on MOS and metallic films; R.C. Dynes, Proc. 16th Intern. Conf. on Low Temperature Physics (Los Angeles, 1981), Physica 108A (1982).
10. E. Abrahams and T.V. Ramakrishnan, J. Non-Crystalline Solids 35, 15 (1980).
11. L.P. Gorkov, A.I. Larkin and D.E. Khmelnitzkii, Soviet Phys. JETP-Lett. 30, 248 (1979).
12. L.P. Gorkov, Soviet Phys.-JETP 36, 1364 (1959).
13. S. Hikami, A.I. Larkin and Y. Nagaoka, Prog. Theor. Phys. 63, 707 (1980).
14. H. Fukuyama, J. Phys. Soc. Jpn. 50, 3407 (1981).
15. Recently B.L. Altshuler, A.G. Aronov, D.E. Khmelnitzkii and A.I. Larkin [preprint] also discussed similar effects. They, however, considered only a part of processes considered in ref.[14].
16. P.A. Lee and T.V. Ramakrishnan, preprint.
17. A. Kawabata, J. Phys. Soc. Jpn 50, 2461 (1981).
18. S. Maekawa and H. Fukuyama, J. Phys. Soc. Jpn. 50, 2516 (1980).
19. L. Van den Dries, C. Van Haesendonck, Y. Bruyneraede and G. Deutscher, Phys. Rev. Lett. 46, 565 (1981) and to be published.

20. See papers presented at 4th Intern. Conf. on Electronic Properties of Two-Dimensional Systems (New London, 1981).
21. T. Ando, Surface Sci. 98, 327 (1980).
22. H. Fukuyama, J. Phys. Soc. Jpn. 49, 649 (1980).
23. H. Fukuyama, Prog. Theor. Phys. Suppl. 69, 220 (1980).
24. H. Fukuyama, J. Phys. Soc. Jpn. 50, 3562 (1981).
25. Y. Kawaguchi and S. Kawaji, J. Phys. Soc. Jpn. 48, 699 (1980).
26. Y. Kawaguchi and S. Kawaji, J. Phys. Soc. Jpn. 49, Suppl. A. 983 (1980).
27. Y. Kawaguchi and S. Kawaji, Proc. Intern. Conf. on Electronic Properties of Two-Dimensional Systems (New London, 1981) to appear in Surface Sci. 114 (1982).
28. D. Vollhardt and P. Wölfle, Phys. Rev. Lett. 45, 842 (1980).
29. W.L. Bloss, L.J. Sham and B. Vinter, Phys. Rev. Lett. 43, 1529 (1980).
30. W.C. McGinnis, M.J. Burns, R.W. Simon, G. Deutscher and P. McChaikin, Proc. 16th Intern Conf. on Low Temperature Physics (Los Angeles, 1981) Physica 107B+C, 5 (1981).
31. T. Kawaguchi and Y. Fujimori, J. Phys. Soc. Jpn. 51, #3 (1982).
32. G. Deutscher and H. Fukuyama, to be published.
33. H. Fukuyama, J. Phys. Soc. Jpn. 51, #4 (1982).
34. F. Komori, S. Kobayashi, Y. Ootsuka and W. Sasaki, J. Phys. Soc. Jpn. 50, 1051 (1981).
35. G. Bergmann, preprints.
36. E. Abrahams, P.W. Anderson, P.A. Lee and T.V. Ramakrishnan, preprint.
37. H. Fukuyama and E. Abrahams, unpublished.
38. G. Bergmann, Phys. Rev. Lett. 43, 1357 (1979).

Localization Versus Superconductivity

S. Maekawa

The Research Institute for Iron, Steel and Other Metals, Tohoku University
Sendai 980, Japan

Abstract

A diagrammatic theory of dirty superconductors is given, which takes into account the effects of localization caused by disorder. It is shown that the mean-field superconducting transition temperature in two dimensions is reduced by the localization.

1. Introduction

ANDERSON[1] and GORKOV[2] have shown in their classic theories of dirty superconductors that the thermodynamic properties such as the transition temperature T_c of a superconductor are not affected by static and non-magnetic disorder. ANDERSON[1] examined exact wave functions of electrons in disordered systems. On the other hand, GORKOV[2] studied the effects of disorder on superconductivity perturbatively. An assumption in their theories was that the interplay between electron-electron interaction and disorder was neglected.

Recently, by extending ANDERSON's argument[1], the interplay in two-dimensional superconductors has been studied, which is a quantum correction related to localization of electrons in disordered systems[3] (this paper is referred to as Ref.3, hereafter). One of the conclusions in Ref.3 is that T_c in the mean-field theory is reduced by the localization. This is because the Coulomb repulsive interaction is enhanced and the density of states of an electron is depressed in dirty systems. This theoretical result is consistent with experimental reports that T_c in a thin film is reduced as the sheet resistance is increased[4-6].

In GORKOV's theory of dirty superconductors[2], the conventional diagrammatic method is applicable and the systematic calculation is possible. Therefore, in this paper we extend the diagrammatic treatment and study the effect of the localization on T_c. We take units $\hbar=k_B=1$.

2. Mean-Field Superconducting Transition

According to Ref.3, we start with the BCS Hamiltonian

$$H = H_0 + g_{BCS}\int d\vec{r}\ [\ \Delta^+\psi_\uparrow(\vec{r})\psi_\downarrow(\vec{r}) + \Delta\psi_\downarrow^+(\vec{r})\psi_\uparrow^+(\vec{r})\ ], \tag{1}$$

where $\psi_\sigma(\vec{r})$ is the electron field operator with spin σ and g_{BCS} is the BCS attractive interaction ($g_{BCS} < 0$). The Hamiltonian H_0 may be written as $H_0 = H_{KE} + H_{imp} + H_{int}$, where H_{KE} means the Hamiltonian of Bloch electrons and H_{imp} the impurity potentials. The Hamiltonian H_{int} includes all kinds of electron-electron interaction except the BCS interaction term given in (1). Here, Δ is the superconducting order parameter given by

$$\Delta = \int d\vec{r}\ \mathrm{Tr}\ e^{-H/T} \psi_{\uparrow}(\vec{r})\psi_{\downarrow}(\vec{r})\ /\ \mathrm{Tr}\ e^{-H/T}. \tag{2}$$

Introducing (1) into (2), we have the following equation for T_c

$$|g_{BCS}|^{-1} = \sum_{k,k'} \int_0^{T^{-1}} du \ \langle T_u\, c_{k\downarrow}{}^{+}(u) c_{-k\uparrow}{}^{+}(u) c_{k'\uparrow}(0) c_{-k'\downarrow}(0) \rangle_{H_0}, \tag{3}$$

where u is the imaginary time, Tu the time ordering operator, and the bracket means the thermal average in the space of H_0. $c_{k\sigma}^{+}$ and $c_{k\sigma}$ are, respectively, the creation and annihilation operators with momentum k and spin σ. The thermodynamic Green's function of an electron is defined by $G(k,\varepsilon_n) = (i\tilde{\varepsilon}_n - \xi_k)^{-1}$ with $\tilde{\varepsilon}_n = \varepsilon_n[1+(2\tau_0|\varepsilon_n|)^{-1}]$, where $\varepsilon_n = 2\pi T(n+1/2)$ with n being integer, ξ_k is the energy of the state with k relative to the Fermi energy ε_F, and τ_0 is the relaxation time of an electron which is expressed by the diagram (Fig. 1). In the figure, the dotted lines with a cross represent impurity potentials (for simplicity, disorder is expressed as impurities), and the solid line is the electron Green's function.

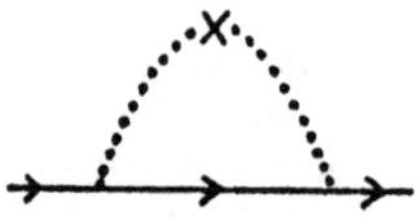

Fig. 1 Self-energy correction due to impurity scattering.

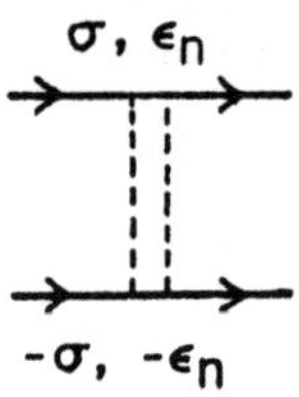

Fig. 2 Vertex correction which originates from Fig. 1. The double broken line denotes the ladder diagram due to impurity scattering between electrons.

The vertex correction of a Cooper pair, which originates from the self-energy given in Fig. 1, is the ladder-type vertex (Fig. 2). Then, the right hand side (r.h.s.) of (3) in this case is written as

$$T \sum_n \sum_k G(k,\varepsilon_n)G(-k,-\varepsilon_n)\ /\ [1 - \frac{1}{2\pi N(0)\tau_0} \sum_{k'} G(k',\varepsilon_n)G(-k',-\varepsilon_n)]. \tag{4}$$

We assume that the density of states of an electron per spin is independent of disorder and is written as N(0) near ε_F. Using the relation

$$\sum_k G(k,\varepsilon_n)G(-k,-\varepsilon_n) = \pi N(0)/|\tilde{\varepsilon}_n|,$$

we have the usual BCS relation

$$|g_{BCS}|^{-1} = T \sum_n \pi N(0)\ /\ |\varepsilon_n|, \tag{5}$$

which is independent of disorder[2].

3. Electron-Electron Interaction

In disordered metals the effects of the electron-electron interaction is enhanced when it is coupled with the diffusive motion of electrons[3,7-10]. In the following, we examine the effects on a Cooper pair propagator. The diagrams of the Hartree-Fock corrections of an electron are given in Fig. 3. In the figure the wavy lines denote the electron-electron interaction. The broken and double broken lines represent the ladder-type vertex corrections due to impurity scattering between electron and hole and between electrons, respectively, both of which may be written as

$$\Gamma(q,\omega_1) = \tau_0^{-1}(Dq^2 + |\omega_1|)^{-1}, \tag{6}$$

with D being the diffusion constant, in the case of $|\omega_1|\tau_0 \ll 1$ and $Dq^2\tau_0 \ll 1$.

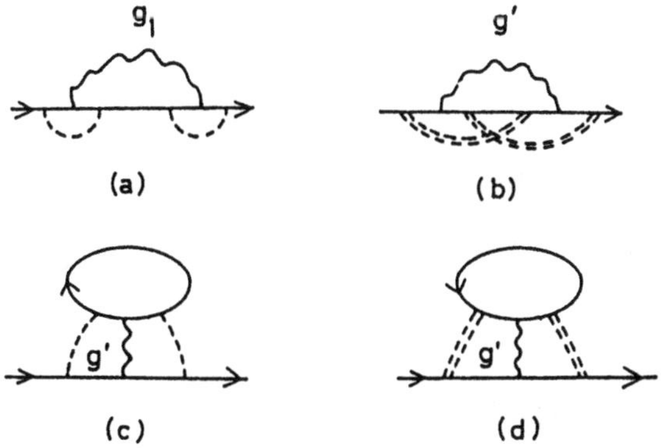

Fig. 3 Self-energy corrections: The processes (a) and (b) are the Fock corrections, and (c) and (d) are the Hartree corrections. The wavy lines denote the interaction g_1 and g'.

The characteristic values of momentum Q and energy ω_1 transferred by the interaction in the process (a) are small, i.e. $|Q| \ll 2k_F$ and $|\omega_1| \ll \tau_0^{-1}$, where k_F is the Fermi momentum, and this interaction is denoted by g_1. On the other hand, the value of Q transferred by the interaction in the processes (b), (c), and (d) can be large, i.e. $|Q| \lesssim 2k_F$ although $|\omega_1| \ll \tau_0^{-1}$. This interaction in the processes is assumed to be equal and is written as g'. We note that g_1 will be strongly repulsive ($g_1<0$) due to the Coulomb interaction, whereas g' will be attractive (g'<0) because the phonon-mediated interaction is dominant at large momentum transfer in superconducting materials; i.e. g' will be close to the BCS coupling constant g_{BCS}.

Let us introduce the correction in Fig. 3(a) into the Cooper pair propagator with energies ε_n and $-\varepsilon_n$ in dirty systems given in Fig. 2. The possible diagrams are given in Fig. 4. We first calculate Fig. 4(a). Since each electron-electron vertex (double broken line) is given by $1/2\tau_0|\varepsilon_n|$, this diagram is expressed as

$$-T \sum_l{}' \sum_q{}' \sum_k \left(\frac{1}{2\tau_0|\varepsilon_n|}\right)^2 g_1[G(k,\varepsilon_n)]^2 G(-k,-\varepsilon_n) G(k-q,\varepsilon_n-\omega_1)[\Gamma(q,\omega_1)]^2, \tag{7}$$

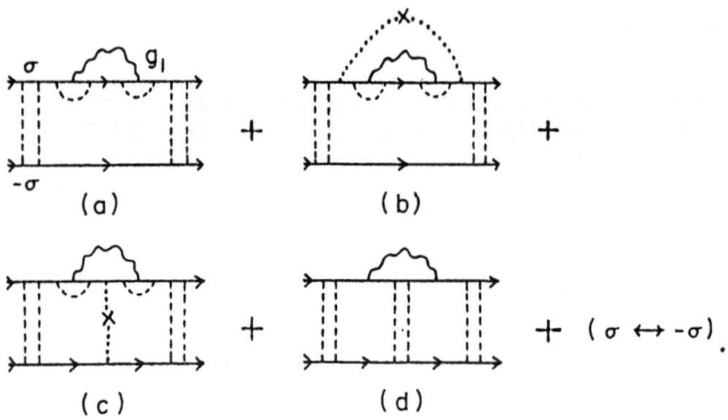

Fig. 4 The Fock corrections of a Cooper pair propagator which originate from Fig. 3(a). The other four diagrams are obtained by changing the spin states ($\sigma\leftrightarrow-\sigma$).

where the primes on the summations denote the conditions $\tau_0^{-1}\gg\omega_1>|\varepsilon_n|$ and $Dq^2\tau_0\ll 1$, respectively. Taking the summation over k, we obtain

$$\text{Fig. 4(a)} = -T\sum_1{}'\sum_q{}'\pi N(0)\tau_0 g_1|\varepsilon_n|^{-2}[\Gamma(q,\omega_1)]^2 \times\left(1+\tau_0|\varepsilon_n|+\tfrac{1}{2}\tau_0|\omega_1|-Dq^2\tau_0\right). \tag{8}$$

The diagrams Fig. 4(b) and (c) are evaluated as

$$\text{Fig. 4(b)} = T\sum_1{}'\sum_q{}'\frac{\pi N(0)\tau_0}{2|\varepsilon_n|^2}g_1[\Gamma(q,\omega_1)]^2\left(1-3Dq^2\tau_0\right), \tag{9}$$

$$\text{Fig. 4(c)} = T\sum_1{}'\sum_q{}'\frac{\pi N(0)\tau_0}{2|\varepsilon_n|^2}g_1[\Gamma(q,\omega_1)]^2\left(1-2Dq^2\tau_0\right). \tag{10}$$

The interaction in the diagram (d) does not change the analytic property of an electron in contrast to that in the diagrams (a), (b), and (c). We evaluate the diagram (d) as

$$\text{Fig. 4(d)} = T\sum_1{}'\sum_q{}'\frac{\pi N(0)\tau_0}{2|\varepsilon_n|^2}g_1\,\Gamma(q,\omega_1). \tag{11}$$

Here, the interaction g_1 transfers the energy $-\omega_1+2\varepsilon_n$, whereas the interaction in the processes (a), (b), and (c) transfers the energy ω_1. Summing up (8) ∿ (11), we have

$$\text{Fig. 4[(a)+(b)+(c)+(d)]} = -T\sum_1{}'\sum_q{}'\pi N(0)\tau_0^2\left\{\frac{g_1(\omega_1)}{|\varepsilon_n|}[\Gamma(q,\omega_1)]^2 + \frac{[g_1(\omega_1)-g_1(2\varepsilon_n-\omega_1)]}{2\tau_0|\varepsilon_n|^2}\Gamma(q,\omega_1)\right\}, \tag{12}$$

where the energies transferred by the interaction g_1 were explicitly expressed. For obtaining (12), we used (6). The corrections are dependent on the dynamical properties of the interaction. This is because the corrections (Fig. 4) occur when the interaction is coupled with the diffusive motion of electrons. Effects of the dynamical interaction will be discussed elsewhere[11]. In the following we assume, for simplicity, the interaction g_1 is instantaneous. Then, the second term in (12) vanishes. Introducing the spin degeneracy, we obtain

$$\text{Fig. 4} = -T \sum_{1}{}' \sum_{q}{}' \frac{2\pi N(0)\tau_0{}^2}{|\varepsilon_n|} g_1[\Gamma(q,\omega_1)]^2. \tag{13}$$

The corrections in Fig. 3(b), (c), and (d) are introduced into the Cooper pair propagator with energies ε_n and $-\varepsilon_n$ in dirty systems given in Fig. 2. The possible diagrams are given in Figs. 5, 6, and 7, respectively. These diagrams are calculated as

$$\text{Fig. 5} = -\text{Fig. 6} = -\text{Fig. 7}$$

$$= -T \sum_{1}{}' \sum_{q}{}' \frac{2\pi N(0)\tau_0{}^2}{|\varepsilon_n|} g'[\Gamma(q,\omega_1)]^2, \tag{14}$$

where the interaction g' is instantaneous.

We consider a thin film whose thickness is less than $\sqrt{D/T}$. In this case, the summation over q in (13) and (14) is restricted within two dimensions. Further taking the summation over ω_1, we obtain the correction caused by Figs. 4 - 7 to the r.h.s. in (3), which is written as R_{HF}, as follows:

$$R_{HF} = -T \sum_n \frac{\pi N(0)}{|\varepsilon_n|} \cdot \frac{(g_1 - 3g')N(0)}{2\pi\varepsilon_F\tau_0} \left[\ln \frac{1}{2\pi T\tau_0} - \psi\left(\frac{1}{2} + \frac{|\varepsilon_n|}{2\pi T}\right)\right], \tag{15}$$

with $|\varepsilon_n|\tau_0 \ll 1$. Here, $\psi(z)$ is the di-gamma function. For obtaining (15), the relations $D = \varepsilon_F\tau_0/m$ and $N(0) = m/2\pi$, m being the effective mass of an electron, were used. As seen in (15), the correction corresponds to the depression of the density of states.

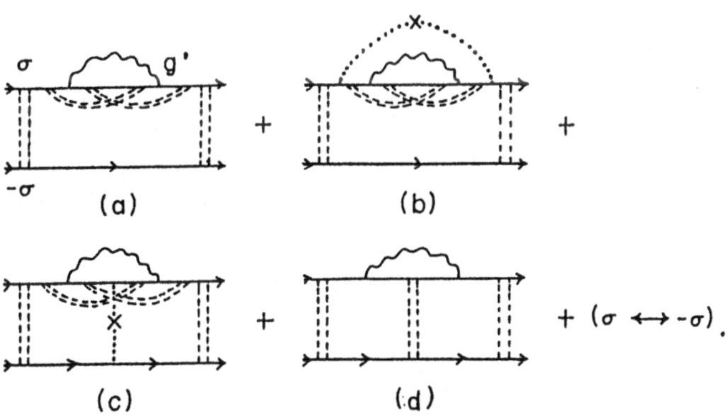

Fig. 5 The Fock corrections of a Cooper pair propagator which originate from Fig. 3(b).

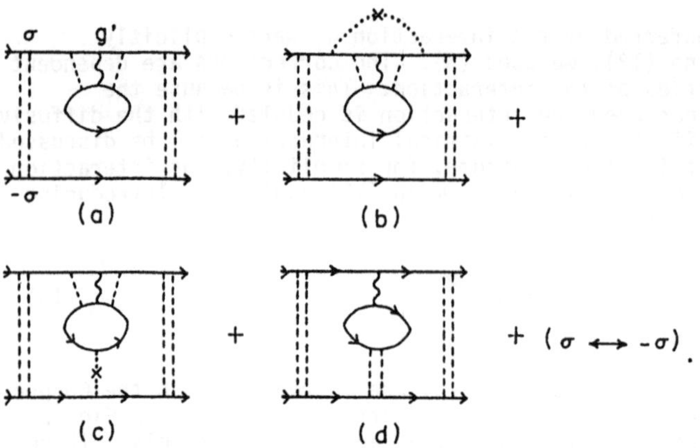

Fig. 6 The Hartree corrections of a Cooper pair propagator which originate from Fig. 3(c).

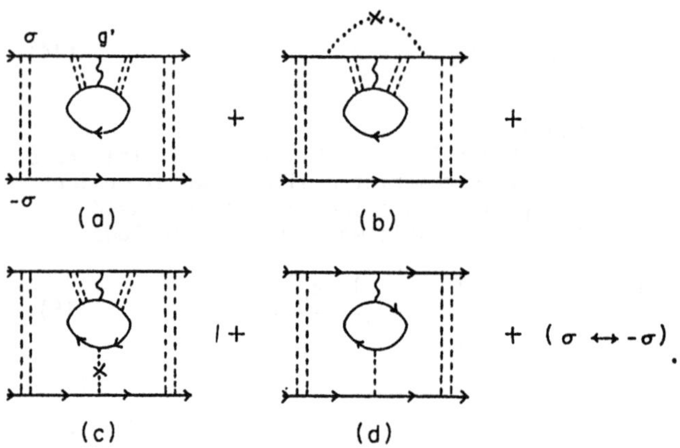

Fig. 7 The Hartree corrections of a Cooper pair propagator which originate from Fig. 3(d).

Let us next examine the vertex corrections caused by the interactions. Figure 8 expresses the possible vertex corrections of a Cooper pair related to the interaction g_1. In Fig. 8(a), the incoming electrons have energies $\pm\varepsilon_n$ and the scattered ones have energies $\pm\varepsilon_{n'}$. The energies ε_n and $\varepsilon_{n'}$ are restricted within the regions $|\varepsilon_n|\tau_0 \ll 1$ and $|\varepsilon_{n'}|\tau_0 \ll 1$. Then, this diagram is written as

$$\text{Fig. 8(a)} = -\sum_k \sum_q{}' g_1 G(k,\varepsilon_n)G(-k,-\varepsilon_n)G(k+q,\varepsilon_{n'})G(-k-q,-\varepsilon_{n'}) \times [\Gamma(q,\varepsilon_n-\varepsilon_{n'})]^2 \frac{1}{4\tau_0{}^2|\varepsilon_n||\varepsilon_{n'}|}, \tag{16}$$

with $\varepsilon_n \cdot \varepsilon_{n'} < 0$. Taking the summation over k, we evaluate (16) as

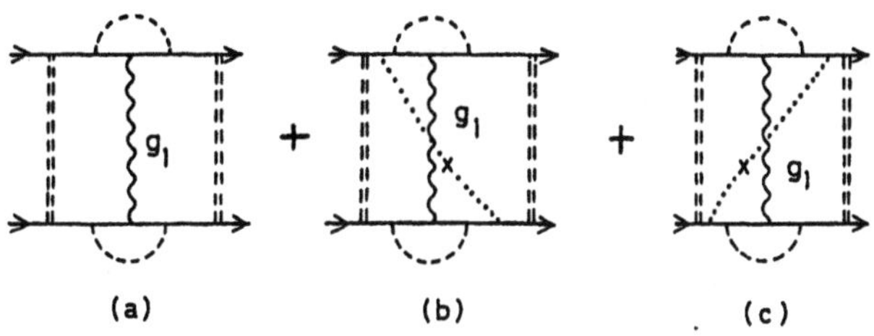

Fig. 8 Vertex corrections due to the interaction g_1.

$$\text{Fig. } 8(a) = -\sum_q{}' \frac{\pi N(0)\tau_0}{|\varepsilon_n||\varepsilon_{n'}|} g_1(1 + \tau_0|\varepsilon_n-\varepsilon_{n'}| - Dq^2\tau_0)$$

$$\times [\Gamma(q,\varepsilon_n-\varepsilon_{n'})]^2 \ . \tag{17}$$

The diagrams Fig. 8(b) and (c) are calculated as

Fig. 8(b) = Fig. 8(c)

$$= \sum_q{}' \frac{\pi N(0)\tau_0}{2|\varepsilon_n||\varepsilon_{n'}|} g_1(1 - 2Dq^2\tau_0)[\Gamma(q,\varepsilon_n-\varepsilon_{n'})]^2 \ , \tag{18}$$

with $\varepsilon_n . \varepsilon_{n'} < 0$, $|\varepsilon_n|\tau_0 << 1$, and $|\varepsilon_{n'}|\tau_0 << 1$. Summing up (17) and (18), we obtain the correction caused by Fig. 8 as

$$\text{Fig. } 8[(a)+(b)+(c)] = -\sum_q{}' \frac{\pi N(0)\tau_0}{|\varepsilon_n||\varepsilon_{n'}|} g_1\Gamma(q,\varepsilon_n-\varepsilon_{n'}) \ , \tag{19}$$

where (6) was used.

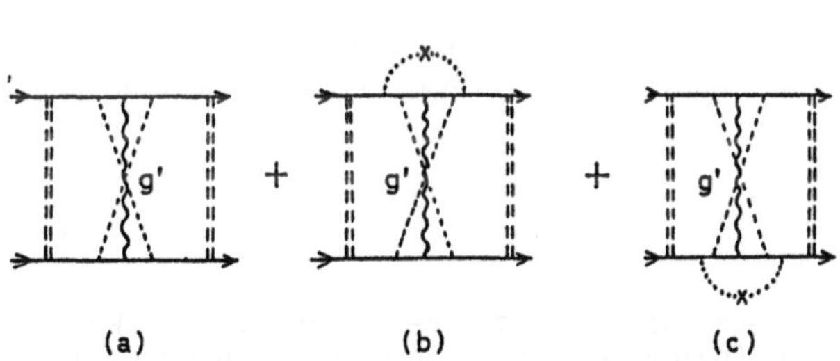

Fig. 9 Vertex corrections due to the interaction g'.

Figure 9 expresses the vertex corrections of a Cooper pair related to the interaction g'. In Fig. 9(a), four Green's functions connected to the interaction g' are absorbed in the electron-electron diffusion propagators because of their definition. Then, this diagram is expressed as

$$\text{Fig. 9(a)} = -\sum_k\sum_q{}' \, g'G(k,\varepsilon_n)G(-k,-\varepsilon_n)G(-k+q,\varepsilon_{n'})G(k-q,-\varepsilon_{n'})$$
$$\times\,[\Gamma(q,\varepsilon_n+\varepsilon_{n'})]^2 \cdot \frac{1}{4\tau_0|\varepsilon_n||\varepsilon_{n'}|}\ , \tag{20}$$

with $\varepsilon_n \cdot \varepsilon_{n'} > 0$, $|\varepsilon_n|\tau_0 \ll 1$, and $|\varepsilon_{n'}|\tau_0 \ll 1$. Taking the summation over k, we obtain

$$\text{Fig. 9(a)} = -\sum_q{}' \frac{\pi N(0)\tau_0}{|\varepsilon_n||\varepsilon_{n'}|} g'(\ 1 + \tau_0|\varepsilon_n+\varepsilon_{n'}| - Dq^2\tau_0\ )$$
$$\times\,[\Gamma(q,\varepsilon_n+\varepsilon_{n'})]^2\ . \tag{21}$$

The diagrams in Fig. 9(b) and (c) are reduced to

$$\text{Fig. 9(b)} = \text{Fig. 9(c)}$$
$$= \sum_q{}' \frac{\pi N(0)\tau_0}{2|\varepsilon_n||\varepsilon_{n'}|} g'(1 - 2Dq^2\tau_0)[\Gamma(q,\varepsilon_n+\varepsilon_{n'})]^2\ , \tag{22}$$

with $\varepsilon_n \cdot \varepsilon_{n'} > 0$. Thus, we find

$$\text{Fig. 9[(a)+(b)+(c)]} = -\sum_q{}' \frac{\pi N(0)\tau_0}{|\varepsilon_n||\varepsilon_{n'}|} g'\Gamma(q,\varepsilon_n+\varepsilon_{n'})\ , \tag{23}$$

with $\varepsilon_n \cdot \varepsilon_{n'} > 0$. When the vertex corrections caused by the interactions g_1 and g' to the r.h.s. in (3) is written as R_v, we obtain the following in two dimensions

$$R_v = -T^2 \sum_{n,n'}{}' \frac{2\pi N(0)}{\varepsilon_n \cdot \varepsilon_{n'}} \frac{(g_1+g')N(0)}{\varepsilon_F\tau_0} \ln \frac{1}{\tau_0(\varepsilon_n+\varepsilon_{n'})}\ , \tag{24}$$

where $\varepsilon_n > 0$, $\varepsilon_{n'} > 0$, and $\tau_0^{-1} \gg (\varepsilon_n+\varepsilon_{n'})$. We take the summations over ε_n and $\varepsilon_{n'}$ in (19) and (23), and retain the most dominant contributions. Then, we have

$$R_v = -\frac{(g_1+g')N(0)}{6\pi\varepsilon_F\tau_0} \left(\ln \frac{1}{T\tau_0}\right)^3 . \tag{25}$$

Here, we assumed that the interactions g_1 and g' are instantaneous. We note that g_1 is essentially the Coulomb repulsive interaction and thus g_1 will overcome the attractive interaction $g' \approx g_{BCS}$. Using (5), (15), and (25), and retaining the dominant contributions, we obtain the equation for T_c

$$\ln \frac{T_c}{T_{c0}} = -\frac{(g_1-3g')N(0)}{4\pi\varepsilon_F\tau_0} \left(\ln \frac{1}{T_c\tau_0}\right)^2 - \frac{(g_1+g')N(0)}{6\pi\varepsilon_F\tau_0} \left(\ln \frac{1}{T_c\tau_0}\right)^3 , \tag{26}$$

where T_{c0} is the transition temperature in the pure system. In the usual superconducting materials, the value of $|g'|N(0)$, which is close to $|g_{BCS}|N(0)$, is much smaller than one. On the other hand, the value of $g_1N(0)$ is of the order of unity. Therefore, g' may be neglected in (26).

4. Discussions

We have obtained that T_c in the mean-field theory is reduced by the localization in two-dimensions. This is because that the Coulomb repulsive interaction is enhanced (the second term in (26)) and the density of states is depressed (the first term in (26)) by multiple scattering of electrons by disorder.

The relation, $(T_c - T_{c0})/T_{c0} = -\beta R$ with $\beta \sim 5\times10^{-4}$ ohm^{-1}, has experimentally been observed in thin films of several materials[4-6]. Since the sheet resistance is given by $R_\square = \pi/e^2\epsilon_F\tau_0$, (26) may be written as

$$\frac{T_c - T_{c0}}{T_{c0}} = -\alpha g_1 N(0) R \left[\left(\ln \frac{1}{T_{c0}\tau_0}\right)^2 + \frac{2}{3}\left(\ln \frac{1}{T_{c0}\tau_0}\right)^3 \right], \tag{27}$$

with $\alpha = 1.2\times10^{-5}$ ohm^{-1}. As seen in (27), T_c is reduced roughly in proportion to $R_\square$, which is consistent with experiment. In thin films observed experimentally[4-6], $R_\square$ is of the order of 10^3 ohm. Therefore, τ_0^{-1} will be $\sim10^2$ K[12]. By use of $T_{c0}\sim5$ K and $g_1N(0)\sim1$, (27) results in the value $\beta \sim 2\times10^{-4}$.

Another interesting feature of (26) is the logarithmic dependence of the correction on temperature. This suggests that the correction becomes strong at low temperatures. A possibility of the reentrance to the normal state due to the localization has been discussed in Ref.3 in connection with experiment by Kobayashi et al.[13].

In this paper, we have studied a perturbational treatment of the effects of the localization on T_c from the metallic limit. Recently, Imry and Strongin[14] have discussed effects of grain size on T_c in granular metals in the context of the metal-non-metal transition picture due to the localization.

Acknowledgements

The work reported has been done in collaboration with Prof. H. Fukuyama. I am thankful to Prof. S. Kobayashi and Dr. T. Kawaguti for informing the experimental results. I am indebted to Prof. M. Tachiki, Dr. H. Ebisawa, and Dr. F. Ohkawa for useful comments.

References

1. P. W. Anderson: J. Phys. Chem. Solids 11, 26 (1959).
2. L. P. Gorkov: Sov. Phys.-JETP 10, 998 (1960).
3. S. Maekawa amd H. Fukuyama: to be published in J. Phys. Soc. Jpn.
4. M. Strongin, R. S. Thompson, O. F. Kammerer, and J. E. Crow: Phys. Rev. 1B, 1078 (1970).
5. D. G. Naugle, R. E. Glover, III, and W. Moormann: Physica 55, 250 (1971).

6. T. Kawaguti and Y. Shibuya: J. Low Temp. Phys. 17, 143 (1974).
7. B. L. Altshuler, A. G. Aronov, and P. A. Lee: Phys. Rev. Lett. 44, 1288 (1980).
8. H. Fukuyama: J. Phys. Soc. Jpn. 48, 2169 (1980) and ibid. 50, 3407 (1981).
9. B. L. Altshuler, D. Khmel'nitzkii, A. I. Larkin, and P. A. Lee: Phys. Rev. B22, 5142 (1980).
10. E. Abrahams, P. W. Anderson, P. A. Lee, and T. V. Ramakrishnan: to be published.
11. H. Ebisawa, S. Maekawa, and H. Fukuyama: unpublished.
12. S. Maekawa and H. Fukuyama: J. Phys. Soc. Jpn. 50, 2516 (1981).
13. S. Kobayashi, Y. Tada, and W. Sasaki: J. Phys. Soc. Jpn. 49, 2075 (1980) and Physica 107B, 129 (1981).
14. Y. Imry and M. Strongin: to be published.

Kondo Effects and Localization in Two Dimensions

F.J. Ohkawa

Institute for Solid State Physics, The University of Tokyo
Roppongi, Minato-ku, Tokyo 106, Japan

Susceptibility, the life-time of particle-particle diffusion propagators and quantum corrections to conductivity due to localized spins are calculated perturbationally with respect to $\lambda = \hbar/2\pi\varepsilon_F\tau$ and exchange coupling constant J. Singular terms proportional to $\lambda J^n(\ln(\hbar/2\pi\tau kT))^2$ are obtained in the susceptibility(n=2) and the life-time(n=3). Another singular term proportional to $\lambda J^3(\hbar/2\pi\tau kT)$ is obtained in the conductivity.

It is discussed that the ground states of localized spins are qualitatively the same as in pure systems if the ensemble average of local density of states given by $g_n = \langle \rho(\varepsilon_1,\mathbf{r})\rho(\varepsilon_2,\mathbf{r})\cdots\rho(\varepsilon_n,\mathbf{r})\rangle$ is non-zero for all n in the limit of equal energies. The self-consistent theory and a physical picture of level repulsion show that g_2 does not vanish even in localized regime.

1. Introduction

Recent scaling argument of localization shows that all the states are localized in two dimensions if only normal impurities, ie spin-independent static impurities, cause disorder [1]. When perturbations dependent on spins such as paramagnetic impurities and spin orbit scatterings are introduced, there appears a possibility in the lowest order scaling argument [2,3] and in the self-consistent treatment [4] that mobility edges may appear. Scatterings are treated in the Born approximation there. Paramagnetic impurities or localized spins are quantum variables. Therefore remarkable effects can be expected if scatterings from localized spins are treated beyond the Born approximation. Such effects are now known as Kondo effects in pure systems [5].

One of the main purposes of this paper is to clarify Kondo effects in weakly localized regime; how Kondo effects are affected by localization. Because perturbational calculations are possible in weakly localized regime, various physical quantities are calculated such as susceptibility, relaxation time, conductivity, and the life-time of particle-particle diffusion propagators.

The life-time of diffusion propagators is very important in the lowest order scaling of localization. A fixed point is determined by the life-time which is related with the amplitude of spin flip scatterings. When it remains finite, localization is destroyed. However it is well known that spin flip scattering tends to zero with decreasing temperatures below Kondo temperature T_K in pure systems. Therefore whether localized spins detroy localization in dimensions lower than or equal to two depends on what is the ground state of localized spins.

The essence of Kondo effects is finite density of states for infinitesimally small excitation energy across the Fermi level. Recent investigations show that envelopes of eigenfunctions go like $\exp(-r/\xi)$ when r is larger than the localization length ξ. According to this simple picture of localization, the system seems to be similar to fine particles whose sizes are about ξ. Moreover the density of states in localized regions is point spectrum even after the thermodynamic limit is reached. Therefore one may doubt whether the ensemble average of n local density of states defined by

$$g_n(\varepsilon_1,\varepsilon_2,\cdots\varepsilon_n) = \langle\rho(\varepsilon_1,\ \mathbf{r})\rho(\varepsilon_2,\ \mathbf{r})\cdots\rho(\varepsilon_n,\ \mathbf{r})\rangle \quad ,$$

with $\rho(\varepsilon,\ \mathbf{r}) = -(1/\pi)\mathrm{Im}\ \langle\mathbf{r}|1/(\varepsilon-H_0+i0)|\mathbf{r}\rangle$ has non-zero value in the limit of equal energies in localized regions; g_n vanishes in the same limit in case of fine particles whose dimensions are much larger than ξ; $\langle A\rangle$ means the ensemble average of A and H_0 is Hamiltonian describing disordered systems. If g_n has non-zero value, it can be expected that Kondo effects do not change qualitatively in localized regions.

Various physical quantities in weakly localized regime are calculated perturbationally in §3 with respect to the degree of disorder $\lambda = \hbar/2\pi\varepsilon_F\tau$ and exchange coupling constant J, after Hamiltonian is introduced in §2. The life-time due to localized spins of particle-particle diffusion propagators is discussed microscopically and it is calculated in §4. In this section, we will see that g_n is important in Kondo effects in localized regime. Thus quantity g_2 is calculated in §5. In §6, summary is presented.

2. Hamiltonian

It is very convenient to use the representation whose bases are eigenstates in the presence of normal impurities: Hamiltonian is divided into two parts as $H=H_0+H_{sd}$ with

$$H_0 = \sum_{\alpha\sigma}\varepsilon_{\alpha\sigma}a^+_{\alpha\sigma}a_{\alpha\sigma} \quad ,$$

and

$$H_{sd} = -\sum_{\mu}\sum_{\mathbf{R}_n}\sum_{\alpha\beta\sigma\sigma'}\phi^*_\alpha(\mathbf{R}_n)\phi_\beta(\mathbf{R}_n)\ a^+_{\alpha\sigma}a_{\beta\sigma'}(J_\mu\sigma^{\sigma\sigma'}_\mu\ S^n_\mu) \quad .$$

Here ε_α and $\phi_\alpha(\ \mathbf{r})$ are the eigenvalues and the eigenfunctions of an electron in the impurity field; J_μ is an exchange coupling constant between electrons and locallized spins ($J_x=J_y=J_\perp$); $\mathbf{R}_n$ is the position of the n-th localized spins, and it is assumed that localized spins are distributed so dilutely that interspin interactions such as RKKY might be neglected.

3. Perturbational Calculations in Weakly Localized Regime

Here various physical quantities are calculated perturbationally with respect to the degree of disorder $\lambda = \hbar/2\pi\varepsilon_F\tau$ and exchange coupling constant J. The results obtained here work well when $\hbar\omega_L \ll kT < \hbar/\tau$, where $\hbar\omega_L = \hbar^2/2m\xi^2\lambda$ is a typical localization energy. Relaxation time of plane wave states due to localized spins is calculated as

$$\hbar/2\tau_s = n_s\pi\rho_0 \sum_\mu (J_\mu/2)^2\{1+4\pi J_\mu\rho_0[\ln(\epsilon_F/kT)+\lambda(\hbar/2\pi\tau kT)\psi^{(1)}(1)+O(\lambda^2)] + O(J^2)\},$$

where ρ_0 is the density of states, $\psi^{(1)}(x)$ is poly-gamma function and n_s is the concentration of localized spins. Conductivity can be expanded as

$$\sigma = (ne^2\tau/m)\ (\ 1 - \ \tau/\tau_s + 2\lambda\ln(\tau/\tau_\epsilon) + \cdots\ )\ ,$$

where τ_ϵ is inelestic relaxation time of particle-particle diffusion propagators, which is introduced in order to make the expansion with respect to J possible; correction to conductivity up to J^3 is simply expressed by relaxation time. In case of constant Coulomb interaction, correction to relaxation time is $\ln T$ [6,7]. In the present case, a more singular term proportional to 1/T is obtained from the same origin; the matrix element of mutual interaction between electrons due to localized spins is singular. In three dimensions, conductivity has a term proportional to $T^{-1/2}$; $T^{1/2}$ dependence in case of constant Coulomb interaction. Because there is a unitarity limit in the cross section of localized spins, this singular term should not remain at low temperatures. As far as perturbation up to J^3 is concerned, one can not obtain any information about the life-time of particle-particle diffusion propagators. This problem is discussed in the next section.

Susceptibility can be expanded as

$$\chi = \chi_0\ \{1 + (1/3)\ \sum_\mu (2J_\mu)^2\ [\ln(\epsilon_F/kT) + \lambda(\ln(\hbar/2\pi\tau kT))^2 + O(\lambda^2)] + O(J^3)\},$$

where $\chi_0 = n_s(g\mu_B)^2 S(S+1)/3kT$ is susceptibility of free spins(S=1/2). It is very interesting to investigate the expansion of susceptibility in detail:

$$\chi(i\omega_\lambda=0) = n_s(g\mu_B)^2 \int_0^\beta d\tau e^{i\omega_\lambda\tau} <\langle T_\tau S_z{}^n(\tau)S_z{}^n(0)\rangle_H>$$

$$= \chi_0\{1-(1/3)\ \sum_\mu (2J_\mu)^2 \int d\epsilon_1 \int d\epsilon_2 g_2(\epsilon_1,\epsilon_2)\frac{f(\epsilon_1)f(-\epsilon_2)}{(\epsilon_1-\epsilon_2)^2}\ [1+\frac{2}{\beta(\epsilon_1-\epsilon_2)}] + \ldots\ \},$$

where $\beta = 1/kT$ and $\langle A\rangle_H = Tr(\exp(-\beta H)A)/Tr(\exp(-\beta H))$. Here it can be seen that the only difference between a pure system and a disordered one lies in the density of states. It is straightforward to prove that this is always the case in every order of exchange coupling constant. When one knows an expression for a pure system, one can obtain easily an expression for a disordered one by replacing energy integrals as

$$\int d\epsilon_1 \cdots \int d\epsilon_n\ \rho_0{}^n \longrightarrow \int d\epsilon_1 \cdots \int d\epsilon_n\ g_n(\epsilon_1,\epsilon_2,\cdots\epsilon_n)\ .$$

The behavior of localized spins depends only on g_n. The singular term obtained perturbationally suggests that g_2 has a logarithmic singularity.

4. Life-Time of Particle-Particle Diffusion Propagators

Dynamical aspects of the localized spins come into play if the scattering is treated in approximations higher than the Born approximation. So far the life-time of particle-particle diffusion propagators, or the inelastic relaxation time $\hbar/\tau_\epsilon$ is introduced phenomenologically in the theory of localization. Here the life-time due to dynamical scatterings by localized

where

$$\hbar/\tau_{\varepsilon 1} = \hbar/\tau_{\varepsilon s} - 2\pi\rho_0(\gamma_0+\gamma_z) \quad , \qquad \hbar/\tau_{\varepsilon 2} = \hbar/\tau_{\varepsilon s} - 2\pi\rho_0(\gamma_0-2\gamma_\perp-\gamma_z) \quad ,$$

and

$$\hbar/\tau_{\varepsilon 3} = \hbar/\tau_{\varepsilon s} - 2\pi\rho_0(\gamma_0+2\gamma_\perp-\gamma_z) \quad .$$

After analytic continuation ($i\varepsilon_n+i\omega_\lambda \rightarrow \varepsilon_F+\hbar\omega+i0$ and $i\varepsilon_n \rightarrow \varepsilon_F-i0$), we can find that spin flip parts give the life-time in the limit of $\omega \rightarrow 0$. For example

$$\hbar/\tau_{\varepsilon s} \rightarrow 2\pi\rho_0(\gamma_0+2\gamma_\perp+\gamma_z) \quad , \qquad \hbar/\tau_{\varepsilon 1} \rightarrow 4\pi\rho_0\gamma_\perp ,$$

$$\hbar/\tau_{\varepsilon 2} \rightarrow 4\pi\rho_0(2\gamma_\perp+\gamma_z) \quad , \text{ and } \qquad \hbar/\tau_{\varepsilon 3} \rightarrow 4\pi\rho_0\gamma_z \, ,$$

where γ_μ is calculated as

$$2\pi\rho_0\gamma_\mu(\varepsilon_F+i0,\varepsilon_F-i0) = 2\pi\rho^{-1}n_s \ll \rho^2(\varepsilon_F, \mathbf{r}) \, |\tau_\mu(\varepsilon_F+i0)|^2 \gg$$

$$\simeq g_2(\varepsilon_F,\varepsilon+\varepsilon_F) \; n_\varepsilon\pi\rho_0(J_\mu/2)^2 \, \{ 1+4J_\mu\pi\,\rho_0[\ln(\varepsilon_F/kT)+ 2\lambda(\ln(\hbar/2\pi\tau kT))^2 +\cdots\}+\cdots .$$

It is well known how $|\tau_\mu|^2$ behaves as a function of frequency or temperature in pure systems. All the components of $|\tau_\mu|^2$ have a maximun around T_K and tend to zero with decreasing temperature and frequency below T_K in case of anisotropy $|J_\perp| > J_z$; they tend monotonically to zero in case of anisotropy $|J_\perp| = J_z$, while only transversal component goes to zero in case of anisotropy $|J_\perp| < J_z$. Life-time $\tau_{\varepsilon 1}$ becomes infinite in the limit of $T \rightarrow 0$ and $\omega \rightarrow 0$ in each case of anisotopy of exchange coupling constant.

Now it becomes clear which diagrams give the life-time of particle-particle diffusion propagators. Including contributions from such terms, we obtain conductivity in weakly localized regime as

$$\sigma = (ne^2\tau/m)\{ 1 - \tau/\tau_s + 2\lambda\ln[(\tau /\tau_{\varepsilon 1})\sqrt{\tau_{\varepsilon 3}/\tau_{\varepsilon 2}} \;]\}.$$

From the results of susceptibility and the life-time in weakly localized regime, it can be seen that localization enhances Kondo effects in such a regime.

5. Local Density of States in Localized Regime

All the conclusions in the previous sections depend on the property of g_n. Therefore g_2 is discussed in order to know how local density of states is affected by localization.

Perturbational calculation in terms of $\lambda = \hbar/2\pi\varepsilon_F\tau$ gives

$$g_2(\varepsilon, \varepsilon+\hbar\omega) = g_1^2(\varepsilon) \; (1 + 2\lambda \, |\ln\omega\tau| + \cdots \;) \quad , \text{ for } \lambda \ll 1,$$

when ω is smaller than $1/\tau$ and larger than $\omega_L=\hbar/2m\xi^2\lambda$. In such a frequency region, singular terms proportional to

$$\lambda \; [\ln(\hbar/2\pi\tau kT)]^2$$

appear in susceptibilty and the life-time of diffusion propagators. The origin of such terms is the logarithmic term in g_2.

When ω is smaller than ω_L, a more precise treatment is necessary. Quantity $g_2(\varepsilon_1,\varepsilon_2)$ can be expressed as

$$g_2(\varepsilon_1,\varepsilon_2) = (2\pi)^{-d} \int d\mathbf{q}\ S(\varepsilon_1,\varepsilon_2;\mathbf{q}) \quad ,$$

with

$$S(\varepsilon_1,\varepsilon_2;\mathbf{q}) = \frac{1}{\Omega} \langle \sum_{\alpha\beta} |\langle\alpha| e^{i\mathbf{q}\cdot\mathbf{r}} |\beta\rangle|^2\ \delta(\varepsilon_1-\varepsilon_\alpha)\delta(\varepsilon_2-\varepsilon_\beta) \rangle ,$$

where Ω is the volume of d-dimensional systems. Function $S(\varepsilon_1,\varepsilon_2;\mathbf{q})$ is related with the polarizability and the conductivity as

$$\chi_e(i\omega_\lambda,\mathbf{q}) = \int d\varepsilon_1 \int d\varepsilon_2 \frac{f(\varepsilon_1)-f(\varepsilon_2)}{i\omega_\lambda-\varepsilon_1+\varepsilon_2} S(\varepsilon_1,\varepsilon_2;\mathbf{q}) \quad ,$$

$$\sigma(i\omega_\lambda,\mathbf{q}) = -i\ (e^2\omega/q^2)\ \chi_e(i\omega_\lambda,\mathbf{q}) \quad ,$$

and

$$\mathrm{Re}\sigma(\omega+i0,\mathbf{q}) = \pi e^2\omega \int d\varepsilon\ [f(\varepsilon)-f(\varepsilon+\hbar\omega)]\ S(\varepsilon,\varepsilon+\hbar\omega;\mathbf{q})/q^2 \quad ,$$

$$\simeq \pi e^2\hbar\omega^2\ S(\varepsilon_F,\varepsilon_F+\hbar\omega;\mathbf{q})/q^2 \quad ,$$

where $f(\varepsilon)$ is the Fermi distribution function. Therefore g_2 is related to conductivity as

$$g_2(\varepsilon_F,\ \varepsilon_F+\hbar\omega) = (2\pi)^{-d} \int d\mathbf{q}\ (q^2 / e^2\hbar\omega^2)\ \mathrm{Re}\sigma(\omega+i0,\ \mathbf{q}) \quad .$$

VOLLHARDT and WÖLFLE [9] investigated localization in the self-consistent scheme for the case of $\lambda \ll 1$. They assumed that the polarizability or the density response function can be written in a diffusion type as

$$\chi_e(\omega+i0,\mathbf{q}) = \rho_0\ D(\omega+i0,\mathbf{q})q^2/(\ D(\omega+i0,\mathbf{q})q^2 - i\hbar\omega\ ) \ ,$$

where the dependence of D on $\mathbf{q}$ is very small. They determined function D in a self-consistent manner. By using their result,

$$S(\varepsilon,\varepsilon+\hbar\omega;\mathbf{q}) = (q^2/\ e^2\hbar\omega^2)\mathrm{Re}\sigma(\omega+i0,\mathbf{q}=0)/(1+q^2\xi^2)^2$$

is obtained. This result is valid for small q such as $\hbar^2q^2/2m < \hbar/\tau$. They showed that $\mathrm{Re}\sigma(\omega+i0,\ \mathbf{q}=0)$ is proportional to ω^2 for small ω such as $\omega < \omega_L$. This leads to a conclusion that $g_2(\varepsilon,\varepsilon+\hbar\omega)$ tends to constant as $\omega \to 0$. According to their theory, the constant is $2/\lambda$. It is generally understood that their theory does not include the effect of level repulsion. When levels repel each other, it seems that $g_2(\varepsilon,\varepsilon+h\omega)$ goes to zero as $\omega \to 0$. Therefore it is very interesting to talk about g_2 by including level repulsion.

MOTT[10] discussed the conductivity by taking level repulsion into account. His discussion predicts that $\mathrm{Re}\sigma(\omega+i0,\mathbf{q}=0)$ is proportional to $\omega^2(\ell n\omega\tau)^{d+1}$ for small ω in d-dimensional systems. It agrees with an exact solution in one dimension for the case of $\lambda \ll 1$ obtained by BEREZINSKI et al.[11-13]. In the present paper, g_2 is discussed by following MOTT.

Let's take localized states in a simple picture of localization following MOTT; each is localized within ξ. He thinks that they are not true eigenstates, but that small overlap integral given by $v(r) \simeq v_0\exp(-r/\xi)$ is left between them where r is the distance of each pair. The effects of such overlap integrals are treated independently in each pair in MOTT's discussion. Two states denoted by a and b, whose energy difference is $\Delta=\varepsilon_b-\varepsilon_a$, form linear combinations;

$$\phi_\alpha(\mathbf{r}) = \cos\theta\phi_a(\mathbf{r}) + \sin\theta\phi_b(\mathbf{r}) \quad \text{and} \quad \phi_\beta(\mathbf{r}) = -\sin\theta\phi_a(\mathbf{r}) + \cos\theta\phi_b(\mathbf{r}) \;,$$

with $\tan 2\theta = v(r_{ab})/\Delta$ and r_{ab} is the distance between states a and b; $r_{ab}=|\mathbf{r}_a-\mathbf{r}_b|$ with $\mathbf{r}_a=\langle a\,|\mathbf{r}|\,a\rangle$ and $\mathbf{r}_b=\langle b\,|\mathbf{r}|\,b\rangle$. An overlap integral makes an energy difference increase from Δ to $\varepsilon=\sqrt{\Delta^2+v^2}$; level repulsion occurs. By neglecting cross terms such as

$$\Omega^{-1}\int d\mathbf{r}\,\phi_a^2(\mathbf{r})\phi_b^2(\mathbf{r}) \simeq 0 \;,$$

a relation is obtained

$$\Omega^{-1}\int d\mathbf{r}\,\phi_\alpha^2(\mathbf{r})\phi_\beta^2(\mathbf{r}) \simeq \cos^2\theta\,\sin^2\theta\,\Omega^{-1}\int d\mathbf{r}\,(\phi_a^4(\mathbf{r})+\phi_b^4(\mathbf{r}))$$

$$\simeq (1/\Omega\xi^d)\,v^2(r_{ab})/(\,v^2(r_{ab})+\Delta^2\,).$$

Summing over "b" for certain "a" in such a condition as $\hbar\omega<\varepsilon<\hbar(\omega+\Delta\omega)$ with an assumption that the states denoted by a or b are distributed independently and homogeneously in space and energy, we obtain

$$\sum_a\Omega^{-1}\int d\mathbf{r}\sum_b \phi_\alpha^2(\mathbf{r})\,\phi_\beta^2(\mathbf{r}) \;\propto\Delta\varepsilon_a\hbar\Delta\omega g_1{}^2(\,\varepsilon_a)\;(\ln(v_0/\hbar\omega))^{d-1}\;.$$

Here summation over b is replaced by integrals over $\mathbf{r}_{ab}$ and Δ as

$$\sum_b \rightarrow g_1(\varepsilon_a)\int d\mathbf{r}_{ab}\int d\Delta \;,$$

with an integration region defined by $\hbar\omega<\varepsilon<\hbar(\omega+\Delta\omega)$ where ε is the energy difference between states α and β as defined above. Summation over "a" is made finally over the states lying between $\varepsilon_a\sim\varepsilon_a+\Delta\varepsilon_a$. Therefore level repulsion gives a result

$$g_2(\varepsilon,\varepsilon+\hbar\omega)\propto(\ln(v_0/\hbar\omega))^{d-1} \;,\quad\text{and}\quad \mathrm{Re}\sigma(\omega+i0;\mathbf{q}=0)\propto(\ln(v_0/\hbar\omega))^{d+1} \;,$$

where the latter was discussed by MOTT. Level repulsion enhances $g_2(\hbar\omega)$ for small ω in dimensions higher than one, and it is different from the simple expectation. An integration over the position of "b"

$$\int d\mathbf{r} \rightarrow \int dr\, r^{d-1} \;,$$

is the origin of the logarithmic contribution $(\ln\omega\tau)^{d-1}$ to g_2 and Re σ. In other words, the origin is that each pair is taken into account independently.

Although it is necessary to discuss further whether this logarithmic singularity remains or not even in the very limit of $\omega\rightarrow 0$, a mechanism called level repulsion gives non-zero $g_2(\omega\rightarrow 0)$ as the self-consistent theory does.

6. Summary

Perturbational calculations have been made in the expansion with respect to the degree of disorder $\lambda=\hbar/2\pi\varepsilon_F\tau$ and exchange coupling constant J. It has been found that there exists a singular term proportional to $\lambda J^2\ln^2(1/t)$ in susceptibility where $t = 2\pi\tau kT/\hbar$; $J^3 \ln^2(1/t)$ in the life-time; a more singular term $\lambda J^3(1/t)$ in conductivity.

It has been shown that the ensemble average of local density of states g_n plays an essential role in Kondo effects in disordered systems. It is discussed based on the self-consistent theory by VOLLHARDT and WÖLFLE and a physical picture of level repulsion by MOTT that $g_2(\varepsilon_1,\varepsilon_2)$ is non-zero in the limit of equal energies. Therefore it can be expected that the ground states of localized spins are qualitatively tha same as in pure systems, when λ is much smaller than one and a typical localization energy is much smaller than Kondo temperature. In such a case, it can be said that localized spins can not destroy localization in dimensions lower than or equal to two unless both temperature and frequency are extremely small. Further investigation is necessary in other cases.

Acknowledgement

The work reported has been done in collaboration with Prof. H. Fukuyama and Prof. K. Yosida.

References

1. E. Abrahams, P.W. Anderson, D.C. Licciardello and T.V. Ramakrishnan, Phys. Rev. Letters 42, 673 (1979).
2. P.A. Lee, J. Non-Crystalline Solid 35 &36, 21 (1980).
3. S. Hikami, A.I. Larkin and Y. Nagaoka, Prog. Theor. Phys. 63, 707 (1980).
4. D. Yoshioka, J. Phys. Soc. Japan 51,(1982).
5. J. Kondo, Solid State Phys. 23, edited by F. Seitz, D. Turnbull and H. Ehrenreich. Academic Press, 1969.
6. B.L. Altshuler, A.G. Aronov and P.A. Lee, Phys. Rev. Letters 44, 1288 (1980).
7. H. Fukuyama, J. Phys. Soc. Japan 48, 2169 (1980).
8. A.A. Abrikosov, Physics 2, 5 (1965).
9. D. Vollhardt and P. Wolfle, Phys. Rev. B22, 4666 (1980).
10. N.F. Mott, Philos. Mag. 22, 7 (1970).
11. V.L. Berezinski, Zh. Eksp. Theor. Fiz. 65, 1251 (1973); Soviet Phys. JETP 38, 620 (1974).
12. A.A. Abrikosov and I.A. Ryzhkin, Zh. Eksp. Theor. Fiz. 71, 1204 (1976); Soviet Phys. JETP 44, 630 (1976).
13. A.A. Abrikosov and I.A. Ryzhkin, Advances in Physics 27, 147 (1978).

Application of the Recent Theory of Anderson Localization to Doped Semiconductors

A. Kawabata

Department of Physics, Gakushuin University
Mejiro, Toshima-ku, Tokyo 171, Japan

1. Introduction

The purpose of this paper is to discuss the electrical conduction in highly doped semiconductors from the point of veiw of the recently developed theory of Anderson localization and other related theories, for instance, those of the effects of electron-electron interaction in disordered systems. Properties of doped semiconductors have been studied for many years [1], and now it is known that in the regions of metallic conduction their fundamental features can be interpreted largely on the basis of electron gas model. The effect of the scattering of electrons by impurities are considered, if necessary, in simple ways, for instance, in Born approximation. The effects of the higher order corrections are believed to be only qualitatively important. However, there are some phenomena which have not been given a clear explanation for a long time. For example, the mechanism of negative magnetoresistance was clarified only recently [2-4] although it was discovered more than 25 years ago [5] and has been known to be one of the general and fundamental properties of doped semiconductors [6]. It is known that the localized spin gives rise to negative magnetoresistance [7], and in some cases the data were analysed on the basis of the localized spin mechanism but no definite conclusion was obtained. Some years ago, KAWAGUCHI et al. [8] observed negative magnetoresistance in a quasi-two-dimensional system (Cs covered Si surface) and found that its magnitude is dependent only on the component of the magnetic field perpendicular to the surface. This fact clearly shows that the orbital motion of electrons is responsible for negative magnetoresistance. Its explanation was given by HIKAMI et al.[9]. They argued that the magnetic field destroys the Anderson localization, which exists in the absence of the magnetic field even in the case of weak disorder [10]. In three-dimensional systems the eigenstates at the Fermi energy are not localized if the random potential is weak enough. Even in the absence of the localization, however, magnetic field suppresses the decrease in the conductivity due to the precursor of the localization. In fact an extension of the theory by HIKAMI et al.to three-dimensional systems could successfully account for various features of negative magnetoresistance in doped semiconductors in metallic conduction regions [3, 4]. Since the theory is an approach from the limit of a good conductor, it is not appropriate to the regions near the metal-insulator transition nor to the regions of hopping conduction. The theory of electron conduction in these regions is not fully developed, at least to such an extent that quantitative calculations of magnetoresistance etc. are possible. Especially in those regions the effects of localized spins should not entirely be ignored, for there is evidence of the existence of a considerable amount of localized spins [1].

Besides the magnetoresistance, the behavior of the conductance at extremely low temperatures cannot be interpreted within a simple theory. In this respect

ALTSHULER and ARONOV [11] have pointed out the importance of the effects of the interference between electron-electron interaction and scattering by random potential, independently of the scaling theory of Anderson localization [10] although it has turned out to give rise to similar effect on the electron conduction. Later the problem was investigated again by ALTSHULER et al.[12]. FUKUYAMA [13] also investigated it independently. In contrast to two-dimensional systems, in three-dimensional systems interaction theory gives a temperature dependence of conductivity qualitatively different from that of one-body localization theory, and it is easier to extract interaction effect from experimental data. In this sense much more study of doped semiconductors, both theoretical and experimental, is required.

In §2 the effects of the localization precursor on the electrical conduction will be discussed within one-body theory. The effects of interaction will be discussed in §3. In §4 the theoretical results will be compared with the experimental data.

2. One-Body Localization Theory

2.1 Single-Valley Semiconductors

As a model of doped semiconductor we consider electron gas interacting with randomly distributed scattering centers which are assumed to be of short range. For the moment we neglect electron-electron interaction. Then one-body Hamiltonian can be written as

$$\mathcal{H} = \frac{1}{2m^*}\left(p + \frac{e}{c} A \right)^2 + \sum_R u\,\delta(r - R) \,. \tag{2.1}$$

The calculations of the conductivity will be done with the use of Feynman gragh method [14]. The graghs of the type shown in Fig.1, which were investigated by LANGER and NEAL [15], are closely related to Anderson localization as

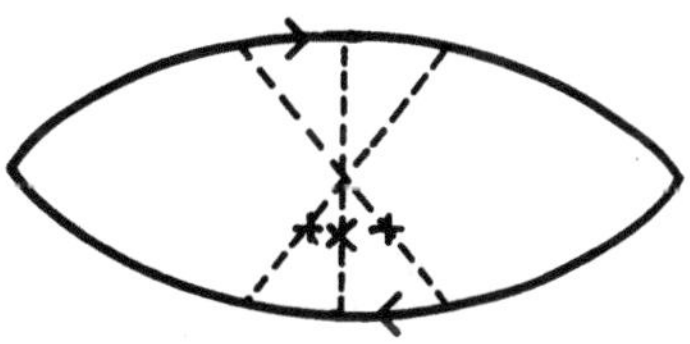

Fig.1 The type of Feynman graphs which give the effect of Anderson localization on the conductivity in the lowest order in $\hbar/\varepsilon_F\tau$. The solid lines and the dotted lines represent one-electron Green's functions and the interaction with the scatterers,respectively

was noted by ABRAHAMS et al. [10] and by GOR'KOV et al. [16]. HIKAMI et al.[9] investigated the effect of magnetic field on the term obtained by summing up the contributions of such graphs. This term represents the effect of the interference of the scattering by different scatterers in the lowest order in $\hbar/\varepsilon_F\tau$, where ε_F is the Fermi energy and τ is the mean scattering time of an electron. In three dimensions this contribution is given by [3]

$$\Delta\sigma(H,T) = -\sigma_0\tau^2\frac{1}{\pi^2\ell^2}\sum_{N=0}^{N_0}\int_{-q_0}^{q_0} dq\,\Gamma_H(N,q,T) \,, \tag{2.2}$$

where σ_0 is the conductivity without the correction given by $ne^2\tau/m^*$, $\ell=\sqrt{c\hbar/eH}$, and Γ_H is the diffusion propagator

$$\Gamma_H(N,q,T) = [2\pi\hbar\nu\tau^2\{4D\ell^{-2}(N + 1/2) + Dq^2 + 1/\tau_\varepsilon\}]^{-1} \,, \tag{2.3}$$

where

$$F = \frac{m^*}{2\pi^2 k_F \hbar^2}\int_0^{2k_F} pv(p)dp \quad , \tag{3.7}$$

and we have dropped q in G's. Then carrying out the summation over ε_n, ε_m and q, we find that

$$(k_BT)^2 \sum_{\varepsilon_n,\varepsilon_m}{}' \sum_q \Gamma(q,\omega_n+\varepsilon_n-\varepsilon_m)\Gamma(q,\varepsilon_n-\varepsilon_m)^2 q_I{}^2$$

$$= \frac{(k_BT)^2}{8\pi(2\pi\hbar\nu\tau^2)^3 D^{5/2}} \sum_{\varepsilon_n>0>\varepsilon_m} \frac{1}{\sqrt{\omega_n+\varepsilon_n-\varepsilon_m}+\sqrt{\varepsilon_n-\varepsilon_m}} \quad . \tag{3.8}$$

Expanding this expression in ω_n and subtracting the contribution of T=0, we obtain

$$\Delta\sigma(T) = -(e^2/2\pi^2\hbar)(k_BT/D\hbar)^{1/2}C_2F \quad , \tag{3.9}$$

where C_2=0.46. The factor C_2 comes out of the expression

$$\frac{\pi^2}{2}(k_BT)^2 \sum_{\varepsilon_n>0>\varepsilon_m} \frac{1}{(\varepsilon_n-\varepsilon_m)^{3/2}} = \frac{\sqrt{2\pi k_BT}}{8}\hbar^{3/2}\sum_{n=1}^{n_0}\frac{1}{\sqrt{n}} \quad , \tag{3.10}$$

with $n_0=\varepsilon_0/2\pi k_BT$, ε_0 being the cut off energy of order $\hbar/\tau$. In the limit T=0, this expression reduces to

$$\frac{\hbar^{3/2}}{8}\int_0^{\varepsilon_0}\frac{d\varepsilon}{\sqrt{\varepsilon}} = \frac{\hbar^{3/2}}{8}(2\pi k_BT)^{1/2}[\sqrt{2}+2(\sqrt{n_0}-\sqrt{n_0-1/2})+\sum_{n=1}^{n_0-1}2(\sqrt{n+1/2}-\sqrt{n-1/2})] \; . \tag{3.11}$$

Subtracting it from (3.10) and letting $n_0\to\infty$, we obtain

$$\frac{(2\pi k_BT)^{1/2}\hbar^{3/2}}{8}[\sum_{n=1}^{\infty}\{\frac{1}{\sqrt{n}}-2(\sqrt{n+1/2}-\sqrt{n-1/2})-\sqrt{2}\;] = -(k_BT)^{1/2}\hbar^{3/2}C_2. \tag{3.12}$$

Summing up the contributions from all the graphs which give the same order of magnitude, we obtain

$$\Delta\sigma_4(T) = (e^2/2\pi^2\hbar)(k_BT/D\hbar)^{1/2}C_2(4/3-3F) \quad . \tag{3.13}$$

As for the factor (4/3 - 3F), 4/3 comes from the exchange type correction considered by ALTSHULER and ARONOV [11] (Their result is twice as large as the present one because of a trivial error), the other exchange type correction associated with the particle-particle propagator gives a contribution proportional to F, and each Hartree correction, associated with particle-particle or particle-hole propagator, gives the factor -2F.

The effect of magnetic field is not simple. As in the case of one-body problem, the particle-particle propagator assumes the form (2.3) and gives rise to magnetoresistance [17]. In addition, Zeeman splitting gives a correction to the propagator, both particle-particle and particle-hole, if it is composed of two electron lines of different spin states [18,19];

$$\Gamma(q,\omega_n) = [2\pi\hbar\nu\tau^2(Dq^2+|\omega_n|-i\omega_z \mathrm{sign}\,\omega_n)]^{-1} \quad , \tag{3.14}$$

ω_z being the Zeeman frequency (if it is a particle-particle propagator, Dq^2 is modified as in (2.3)). The general expression for the magnetoconductance is complex, and below we show some special cases. The effect of Zeeman splitting shows up only when $2\pi k_B T < \hbar\omega_z$. The orbital effect shows up when $\hbar D\ell^{-2} > 2\pi k_B T$ and in typical cases this condition is fulfilled at higher temperature or lower magnetic field. When the Zeeman splitting is negligible, the magnetoconductance is given by

$$\Delta\sigma_3(H,T) = \frac{3e^2 F}{4\pi\hbar^2 \ell} k_B T \sqrt{a} \sum_{n=1}^{\infty} \omega_n \sum_{N=0}^{\infty} [\{a(N+\tfrac{1}{2})+\omega_n\}^{-5/2} + \frac{2}{3a}[\{a(N+1)+\omega_n\}^{-3/2} - (aN+\omega_n)^{-3/2}]], \qquad (3.15)$$

where $a=4D\ell^{-2}$. When $a \ll k_B T/\hbar$, we expand the right hand side in a and obtain

$$\Delta\sigma_3(H,T) = -\frac{e^2}{2\pi^2\hbar\ell} F \frac{5}{64} \zeta(\tfrac{5}{2}) \left(\frac{a\hbar}{2\pi k_B T}\right)^{3/2}, \qquad (3.16)$$

$\zeta(5/2)(=1.34)$ being the Riemann's zeta function. At T=0, (3.15) reduces to

$$\Delta\sigma_3(H,0) = -(e^2/2\pi^2\hbar\ell)F\cdot F(0) = -0.92F\sqrt{H} \text{ mho/cm} \quad (H \text{ in kOe}), \qquad (3.17)$$

F(0) being defined by (2.6). The expression for $\Delta\sigma_3(H,0)$ including Zeeman splitting is

$$\Delta\sigma_3(H,0) = -(e^2/2\pi^2\hbar\ell)F\cdot[F_\lambda + \sqrt{\lambda/2}\,], \qquad (3.18)$$

where

$$\lambda = \omega_z/4D\ell^{-2} = 3\hbar|g|m^*/16\varepsilon_F\tau m_0, \qquad (3.19)$$

and

$$F_\lambda = \sum_{N=0}^{\infty} [\frac{2}{\sqrt{N+1}+\sqrt{N}} - \{\frac{1}{2}\frac{((N+1/2)^2+\lambda^2)^{1/2}+(N+1/2)}{(N+1/2)^2+\lambda^2}\}^{1/2}], \qquad (3.20)$$

with m_0 the electron mass in vacuum and g the g-factor of the spin. In (3.18) F_λ represents the contribution from the Hartree term associated with the particle-particle propagator of electrons of different spin states, and $\sqrt{\lambda/2}$ comes out of the contribution of Zeeman splitting. The pure orbital contributions from the Hartree and exchange terms cancel exactly. As is seen from (3.19), λ is very small and the Zeeman splitting effect is negligible unless $\varepsilon_F\tau/\hbar$ is not very large and $|g|$ is larger than m_0/m^*.

3.2 Many-Valley Semiconductors

In the case of many-body effect the contributions from each valley can not be treated separately even if the intervalley scattering is negligible, for the electrons in different valleys interact via Coulomb interaction. The multiplicity of the valleys gives rise to little difficulty if the effective masses of the valleys are isotropic or have the same anisotropy. In the cases like Ge or Si, the anisotropy shows up only in the evaluation of the factor (4/3-3F) in (3.13). Consider the contribution from the gragh of Fig.3; in the case of Ge, the diffusion propagator composed of two electron lines in different valleys does not exhibit the singularity at $q=\omega_n=0$ because of mismatching of the Fermi surface. In the case of Si only the electrons in two valleys directed in the same way contribute to the singular diffusion propagator as well as those in the same valley. Thus, in (3.2), except for the last factor, which is essentially F, the problem can be reduced to isotropic case as in one-body problem because the G's and Γ's involved are those of the same valley or valleys with the same anisotropy. Thus at H=0, the total correction is given by

$$\Delta\sigma_4(T) = S_0(e^2/2\pi^2\hbar)(k_BT/D\hbar)^{1/2}C_2(A/n_\nu) , \qquad (3.21)$$

where A reduces to $4/3+F(1-4n_\nu')$ in isotropic case, n_ν is the number of the valleys, and n_ν' is the number of the valleys of common energy surface (e.g., $n_\nu'=1$ for Ge and $n_\nu'=2$ for Si). Here F is defined by

$$F = (n_\nu m_a/2\pi^2\hbar k_F)(1/8\pi^2)\int d^3p d^3k \delta(\varepsilon(p)-\varepsilon_F)\delta(\varepsilon(k)-\varepsilon_F)v(p-k), \qquad (3.22)$$

where $\varepsilon(p)$ is the one-electron energy and k_F is defined by $\hbar^2k_F^2/2m_a=\varepsilon_F$. Roughly speaking v(p-k) is smaller than that of single-valley case by a factor n_ν because all valleys contribute to the screening, and hence within Thomas-Fermi approximation F reduces to 1 if the screening length is short enough. Thus, the contribution to $\Delta\sigma_4(T)$ per valley is reduced by a factor $1/n_\nu$ compared with single-valley case. On the other hand, contribution of one-body effect is approximately proportional to n_ν, and hence many-body effect is less important in many-valley semiconductors than in single-valley cases.

As for the magnetoconductance, we give only the result of T=0,

$$\Delta\sigma_3(H,0) = -(e^2/2\pi^2\hbar\ell)\sum_i\alpha_i(m_a/m_{ci})^{1/2}(F/n_\nu)[n_\nu'(F_{\lambda_i}+(\lambda_i/2)^{1/2}) +(n_\nu'-1)F(0)] , \qquad (3.23)$$

λ_i being given by

$$\lambda_i = 3\hbar|g_i|m_{ci}/16\varepsilon_F m_0\tau , \qquad (3.24)$$

where $g_i=e_H\cdot g\cdot e_H$ with g the g-tensor in the ith valley and e_H is the unit vector along the magnetic field.

4. Comparison with Experiments

There are a lot of data on the magnetoresistance and the temperature dependence of the conductivity. In the theoretical calculations in the preceeding sections we have neglected the effects of spin-orbit interaction, localized spin and intervalley scattering, and we have to be careful about these effects in comparing the data with the theoretical results.

4.1 Single-Valley Semiconductors

There are various III-V compound semiconductors of which the minimum of the conduction band is at the center of the Brillouin zone with approximately isotropic effective mass. Among them GaAs and InP are suitable for our purpose, for spin-orbit coupling in the conduction band should be very small because of the small spin-orbit splitting in the valence band compared with the band gap. A detailed measurement of the magnetoresistance of GaAs was carried out by EMELI'ANENKO and NASLEDOV [20]. In Fig.4, we show the results of the experiment. In Table 1 we show the parameters characterizing the samples. The data on the magnetoresistance used to be represented in the form

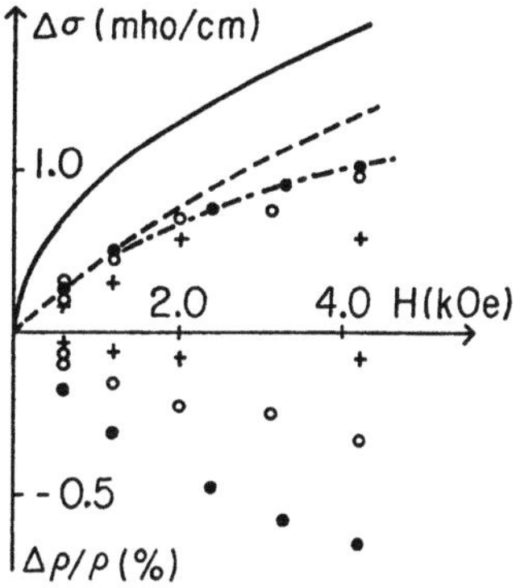

Fig.4 Magnetoresistance ($\Delta\rho/\rho$) and magnetoconductance ($\Delta\sigma$) of GaAs at 4.2K on three samples, NO.2(•), No.3(o), No.4(+). The lines show the theoretical results; one-body theory at 0K (———), one-body theory with $\tau_\varepsilon=1.0\times10^{-11}$ sec(– – – –), many-body effect included(—·—·—) (for No.2).

Table 1 Carrier density n, conductivity σ_0 at 1.5K, and other parameters of n-type GaAs samples (F is defined by (3.7))

Sample	$n(10^{18}cm^{-3})$	σ_0(mho/cm)	$2\pi\varepsilon_F\tau/h$	F
No. 2	0.5	156	25	0.43
No. 3	0.75	298	40	0.40
No. 4	3.1	690	59	0.31

$\Delta\rho/\rho$, e.g., its ratio to the resistance at H=0. In Fig.4 we find that $\Delta\sigma$ is largely independent of the sample for H<3kOe in contrast to $\Delta\rho/\rho$. This behavior is consistent with the thoeretical prediction (2.9). If we apply the one-body localization theory at T=0, it gives value about twice as large. The best fit for the sample No.2 for H≤2kOe has been obtained by putting $\tau_\varepsilon=1.0\times10^{-11}$ sec in (2.6). In this region the many-body effect is not important and we find that the fitting for 2kOe<H<4kOe is improved if we take account of many-body effect. The effect of Zeeman splitting was neglected because λ was estimated to be 0.3×10^{-3}. The value of τ_ε assumed here is reasonable if we consider electron-phonon interaction. Electron-electron interaction gives $\tau_\varepsilon\simeq10^{-10}$sec. The enhancement of the relaxation due to randomness discussed by SCHMID [21], by ALTSHULER and ARONOV [22] and also by ABRAHAMS et al. [23] is not effective since the factor of the enhancement is of order $1/(k_BT\tau/\hbar)^{1/2}\varepsilon_F\tau$ and is smaller than 1.

4.2 Many-Valley Semiconductors

As a typical many-valley semiconductor we consider Ge. First we consider the anisotropy of the magnetoresistance, which is characteristic to many-valley semiconductors with anisotropic effective mass. The anisotropy was measured by SASAKI [24] on a n-type sample of the carrier density $3.0\times10^{17}cm^{-3}$ at T=4.2K and H=1kOe (Fig.5). In this region $\Delta\rho$ is proportional to H^2, and hence the anisotropy is determined by S_2 in (2.26) (it is the case even if the many-body effect is considered although it is not important in this region). Here we will be concerned only with the anisotropy and hence we put

$$\Delta\rho/\rho = - AS_2 \qquad (4.1)$$

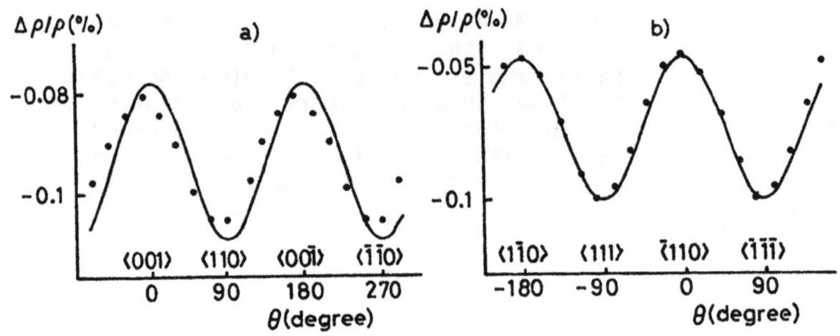

Fig.5 Anisotropy of the magnetoresistance of n-type Ge ($n=3.0\times10^{17}cm^{-3}$) at 4.2K and 1kOe. The lines and the points show the theory and the experiment, respectively.

with A an adjustable parameter. The experiment was done on two kind of geometries; the magnetic field was rotated in the plane perpendiclar to a) $\langle\bar{1}10\rangle$, and b) $\langle11\bar{2}\rangle$. The current is along $\langle\bar{1}10\rangle$ in both cases. In the case of a), by θ we represent the angle between the magnetic field and ⟨001⟩. Then from (2.15) and (2.24) it is easy to see that

$$S_2 = 27.9(1 - 0.29\cos^2\theta) \; . \qquad (4.2)$$

In the case b), the angle between the magnetic field and $\langle\bar{1}10\rangle$ is taken to be θ, and we obtain

$$S_2 = 25.3(1 - 0.53\cos^2\theta). \qquad (4.3)$$

With A=0.0039%, the theory remarkably well reproduces the experiment as is shown in Fig.5. This good agreement is not accidental, for the function S_x can be derived from a very general argument under the assumptions 1) the orbital motion of electrons is responsible for the magnetoresistance: 2) the scatterers are of short range type: 3) the contribution from each valley is additive: 4) anisotropy is absent if the effective mass is isotropic. Namely, by a transformation we can reduce the problem to isotropic case with H replaced by $H'=Hm_a/m_{ci}$. Then, the contribution from ith valley can be written in the form

$$\Delta\sigma_i = \alpha_i f(Hm_a/m_{ci}) \; , \qquad (4.4)$$

f being some function. Then in the regions where $\Delta\sigma\propto H^x$, we obtain

$$\Delta\sigma_i \propto \alpha_i (m_a/m_{ci})^x H^x \; , \qquad (4.5)$$

and we get $S_x H^x$ if it is summed over the valleys.

This experiment on the anisotropy is very important, because it provides firm evidence that the orbital motion is responsible for the magnetoresistance. If it is caused as a result of interaction of electron spins with magnetic field, the anisotropy should show up through that of g-factor. From an argument as above we can show that the anisotropy is determined by S'_x defined by

$$S'_x = \sum_i \alpha_i (g_i/g_a)^x \quad , \qquad (4.6)$$

where g_i was defined in the last part of §3, and $g_a=(g_t{}^2 g_\ell)^{1/3}$. In the present case, from g_t=0.87 and g_ℓ=1.9 [24] it follows that

$$S_2' = 7.83(1 + 0.19\cos^2\theta) \quad (a) \quad ,$$
$$= 5.71(1 + 0.44\cos^2\theta) \quad (b) \quad . \qquad (4.7)$$

Comparing it with (4.2) and (4.3) we find that the positions of the maximum and the minimun have to be interchanged if the spin mechanism is dominant.

Dependence of magnetoresistance on the magnitude of the magnetic field was carefully measured by ROTH et al. [26]. They analysed the data assuming the form

$$\Delta\rho/\rho = aH^c + bH^2 \; , \qquad (4.8)$$

where the 2nd term is supposed to represent the normal magnetoresistance. A part of their results are shown in Table 2. We find that the exponent c tends to 1/2 at low temperature for the samples of higher carrier concentration. This behavior is consistent with the square root dependence given by (2.25) and (3.23). On the other hand, the factor a is smaller than the theoretical value by an order of magnitude (in Ref.26 the value of ρ and the directions of the current and the magnetic field are not given but the discrepancy can not be explained in terms of those ambiguities). It may be explained by considering intervalley scattering. If the intervalley scattering is strong enough magnetoconductance is reduced to that of single-valley system with istropic effective mass [27,28]. In principle the strength of intervalley scattering can be estimated from anisotropy measurements.

Table 2 The exponent c of the field dependence of the magnetoresistance of n-type Ge, N_D being the donor concentration.

$N_D(10^{19}cm^{-3})$	c 4.2K	c 1.8K
0.13	0.98	0.72
0.30	0.75	0.57
0.88	0.62	0.51
1.23	0.63	0.51

Measurement of magnetoresistance of n-type Ge down to 10mK has been carried out by OOTUKA et al.[29]. In the case of carrier density $3.3\times10^{17}cm^{-3}$, for example, they found that the magnetoresistance becomes positive below 0.1K∿1K (dependent on the magnitude of H; see Fig.6). Putting $\Delta\sigma=CH^x$, the best fit was obtained with C=-2.2mho/cm and x=0.53 for T=0.1K. . The theoretical value can be obtained from (2.25) and (3.23):

$$\Delta\sigma_1(H,0) = 8.5\sqrt{H}\ \text{mho/cm} \; , \quad (\text{H in kOe}) \qquad (4.9)$$

$$\Delta\sigma_3(H,0) = -\ 3.0\sqrt{H}\ \text{mho/cm} \; , (\text{H in kOe}) \qquad (4.10)$$

where the current is along <$\bar{1}$10> and the magnetic field is along <111>. We

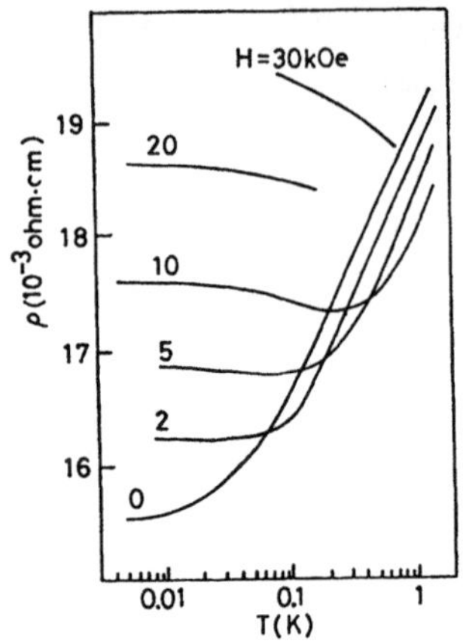

Fig.6 Resistivity of n-type Ge ($n=3.3\times10^{17}cm^{-3}$) vs. logT for various values of magnetic field.

have evaluated F using RPA screened Coulomb interaction with full anisotropy and it was found to be nearly 1. From the above we find that in order to reproduce the experimental value we have to assume that the contribution of the many-body effect is 3.5 times as large as the present calculation (provided that the contribution of the one-body effect is not changed). This enhancement factor 3.5 is consistent with the temperature dependence of σ at H=0 to be discussed below although there is no way of justifying it at present. Another way of explaining the experiment is to assume strong spin-orbit coupling. In Ref.9 it is shown that in the presence of strong spin-orbit coupling the sign of magnetoresistance changes with reduction of its magnitude by a factor 2 (this is the case also in three-dimensional systems). However, we can not explain the whole feature of the conductivity in a consistent way with this assumption. Moreover a simialr behavior exhibited by P doped Si [30] can not be explained in this way , for we can not expect strong spin-orbit coupling in the system composed of such light elements.

Detailed analysis of the temperature dependence of the conductivity of Ge at H=0 was made by THOMAS et al.[31]. If we assume electron-electron scattering for τ_ε, i.e., if $1/\tau_\varepsilon \propto T^2$, we may put (cf. (2.23) and (3.21))

$$\Delta\sigma(0,T) = \Delta\sigma_2(T) + \Delta\sigma_4(T) = BT + mT^{1/2} . \qquad (4.11)$$

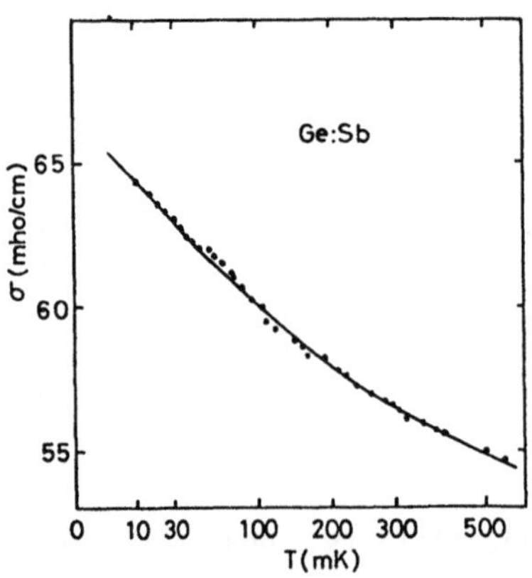

Fig.7 Conductivity of n-type Ge ($n=3.3\times10^{17}cm^{-3}$) vs. $T^{1/2}$ at zero magnetic field.

The values of B and m are determined to get the best fit. Figure 7 shows $\Delta\sigma(0,T)$ of the sample of $n=3.3\times10^{17}cm^{-3}$. Table 3 shows the results of the analysis for various samples. The theoretical values of m are listed in the forms of the ratios of the experimental values to them. The values of B are shown in the forms of $\tilde{c}$, which will be defined below. In calculating m using (3.21), we evaluated F within RPA with full anisotropy. The calculations were done also for isotropic model with effective mass m_a and it was found that in both cases F is nearly 1 (within 10%) for all the samples. The same situation can be expected for the factor 4/3, and in (3.21) we have put A=4/3-3. We used the isotropic model also in evaluating $1/\tau_\varepsilon$. Within Born approximation assuming RPA screened Coulomb interaction, we obtain [32]

$$1/\tau_\varepsilon = cT^2 , \qquad (4.12)$$

where

$$c = 4(\pi/4)^3(3\pi)^{-1/2}(n_\nu/3\pi^2)^{1/3}(m_a k_B{}^2/\hbar^2 n^{5/6} a_B{}^{1/2}) , \qquad (4.13)$$

with a_B the effective Bohr radius. Thus the parameter $\tilde{c}=c\tilde{n}^{5/6}$, $\tilde{n}$ being n in the unit of $10^{18}cm^{-3}$, is independent of n and takes the value

$$\tilde{c} = 0.17\times10^{10}sec^{-1}K^{-2} . \qquad (4.14)$$

Table 3 The coefficient m of the $T^{1/2}$ term of the conductivity of n-type Ge samples with various carrier concentration. $\tilde{c}$ is related to T-linear term (cf. (4.11) to (4.15))

$n(10^{18}cm^{-3})$	σ_0(mho/cm)	$2\pi\varepsilon_F\tau/h$	$m(mho cm^{-1}K^{-1/2})$	m/m_{theory}	$\tilde{c}(10^{10}K^{-2}sec^{-1})$
0.21	32	1.4	-21	2.7	0.88
0.26	40	1.6	-16	2.4	0.44
0.33	67	2.5	-18	3.3	0.70
0.48	65	2.2	-12	2.1	0.25
0.64	114	3.4	- 9	2.1	0.32

Experimentally $\tilde{c}$ is obtained from B as

$$\tilde{c} = D\tilde{n}^{5/6}[B/(S_0 e^2/2\pi^2\hbar)]^2 , \qquad (4.15)$$

where diffusion coefficient D is defined by

$$\sigma_0 = 2e^2 S_0 D\nu , \qquad (4.16)$$

with ν the density of state per valley,

$$\nu = (3\pi^2 n/n_\nu)^{1/3} m_a/2\pi^2\hbar^2 . \qquad (4.17)$$

From Table 3 we find that the values of $\tilde{c}$ obtained in this way are in agreement with the theory in the order of magnitude. If we consider the effect of randomness on $1/\tau_\varepsilon$, it should behave like $T^{3/2}$[21,22,23], but it is rather difficult to distinguish $T^{3/4}$ law from T-linear law within the present experimental precision. The tendency of the large experimental values of $\tilde{c}$ seems to indicate the enhancement of $1/\tau_\varepsilon$ due to randomness. As for the $\sqrt{T}$ term, m are 2∿3 times as large as the theoretical values. In the case of $n=3.3\times10^{17}cm^{-3}$, this ratio is 3.3 and hence it is consistent with the factor 3.5 of the enhancement of many-body effect in the case of magnetoresistance.

5. Effects of Spin-Orbit Coupling

HIKAMI, LARKIN and NAGAOKA [9] discussed the effects of the spin-orbit coupling and localized spin on the localization. For two-dimensional systems they showed that if the spin flip scattering time τ_{so} due to spin-orbit coupling is shorter than τ_ε, the sign of the magnetoresistance is reversed with a reduction of the magnitude by a factor 2. It is easy to see this is the case for three-dimensional systems. ROTH et al. [26] observed positive magnetoresistance on p-type Ge which is proportional to $H^{0.44\sim0.46}$ at T=1.8K. Since the spin-orbit coupling is expected to be strong in the valence band, this experiment may be explained in terms of the effect mentioned above. SUGIYAMA and KOBAYASHI [33] also observed that and in addition they found that the magnetoresistance becomes negative under uniaxial stress. This effect may be interpreted in the following way [4]; under the stress the four fold degeneracy of the valence band is lifted. From the argument of time reversal symmentry we can show that the intraband spin-flip scattering is much less frequent than the interband one [34], and then τ_{so} increases very much under the stress. However, in this experiment the samples are not metallic enough. Moreover it is not clear how the complexity of the valence band affects the magnetoresistance, and a definite conclusion is not yet obtained.

6. Summaries

We have seen that some important properties of the electron conduction in doped semiconductors can be interpreted in terms of the theory of Anderson localization and other related theories developed recently. A simple model of doped semiconductors, electron gas interacting with randomly distributed impurities, can account for the magnetoresistance, which has been a puzzle for a long time, and the temperature dependence of the conductivity in low temperature regions, at least qualitatively, if we properly take account of the effects of the interference of the scattering of electrons by different impurities and of electron-electron interaction. However, there are some properties for which quantitative agreements between the theory and the experiment were not obtained. For example, we have to assume an enhancement of many-body effect to explain the behavior of Ge at T≲1K, and further investigations, both theoretical and experimental, are required. In this respect, the measurement of the anisotropy of magnetoresistance is very important, for it gives important information on the roles of spins (including localized ones), intervalley scattering, etc.

References

1. W. Sasaki: Proc. Intern. Conf. Semiconductors, Kyoto, J. Phys. Soc. Japan 49 Supple. A, 31 (1980).
2. A. Kawabata: Solid State Commun. 34, 432 (1980).
3. A. Kawabata: J. Phys. Soc. Japan 49, 628 (1980).

4. A. Kawabata: Proc. Intern. Conf. Semiconductors, Kyoto, J. Phys. Soc. Japan 49 Supple. A, 375 (1980).
5. H. Fritzsche and K. Lark-Horovitz: Phys. Rev. 99, 400 (1955).
6. W. Sasaki: Proc. Intern. Conf. Semiconductors, kyoto, J. Phys. Soc. Japan 21 Supple., 543 (1966).
7. K. Yosida: Phys. Rev. 107, 396 (1957).
8. Y. Kawaguchi, H. Kitahara and S. Kawaji: Surface Sci. 73, 520 (1978).
9. S. Hikami, A.I. Larkin and Y. Nagaoka: Prog. Theor. Phys. Letters 63, 707 (1980).
10. E. Abrahams, P.W. Anderson, D.C. Licciardello and T.V. Ramakrishnan: Phys. Rev. Lett. 42, 673 (1979).
11. B.L. Altshuler and A.G. Aronov: Zh. Eksp. Teor. Fiz. 77, 2028 (1979) (Sov. Phys. JETP 50, 968 (1979)).
12. B.L. Altshuler, D.E. Khmelnitskii, A.I. Larkin and P.A. Lee: Phys. Rev. B 22, 5142 (1980).
13. H. Fukuyama: J. Phys. Soc. Japan 48, 2169 (1980).
14. A.A. Abrikosov, L.P. Gor'kov and I.E. Dzyaloshinskii: Methods of Quantum Field theory in Statistical Physics (Pergamon Press, London, 1963).
15. J.S. Langer and T. Neal: Phys. Rev. Lett. 16, 984 (1966).
16. L.P. Gor'kov, A.I. Larkin and D.E. Khmelnitskii: ZhETP Lett. 30, 248 (1979).
17. H. Fukuyama: J. Phys. Soc. Japan 50, 3407 (1980).
18. A. Kawabata: J.Phys. Soc. Japan 50, 2461 (1980).
19. P.A. Lee and T.V. Ramakrishnan: preprint.
20. O.V. Emeli'anenko and D.N. Nasledov: Zh. Tekh. Fiz. 28, 117 (1958) (Sov. Phys. Tech. Phys. 3, 1094 (1959)).
21. A. Schmid: Z. Physik 271, 251 (1974).
22. B.L. Altshuler and A.G. Aronov: Pis'ma Zh. Eksp. Teor. Fiz. 30, 514 (1979)(Sov. Phys. JETP Lett. 30, 482 (1979).
23. E. Abrahams, P.W. Anderson, P.A. Lee and T.V. Ramakrishnan: preprint.
24. W. Sasaki: J. Phys. Soc. Japan 20, 825 (1965).
25. Y. Yafet: Solid State Physics, eds. F. Seitz and D. Turnbull (Academic Press, New York 1963) Vol. 14, p. 54.
26. H. Roth, W.D. Straub, W. Bernard and J.E. Mulhern, Jr.: Phys. Rev. Lett. 11, 328 (1963).
27. H. Fukuyama: J. Phys. Soc. Japan 49, 649 (1980).
28. B.L. Altshuler, D.E. Khmelnitskii, A.G. Aronov and A.I. Larkin: preprint.
29. Y. Ootuka, S. Kobayashi, S. Ikehata, W. Sasaki and J. Kondo: Solid State Commun. 30, 169 (1979).
30. Y. Ootuka : private communication.
31. G.A. Thomas, A. Kawabata, Y. Ootuka, S. Katsumoto, S. Kobayashi and W. Sasaki: to be published.
32. J.J. Quinn and R.A. Ferrell: Phys. Rev. 112, 812 (1958).
33. K. Sugiyama and A. Kobayashi: J. Phys. Soc. Japan 18, 163 (1963).
34. p. 87 of Ref. [25].

Dielectric and Optical Anomalies Near the Anderson Metal-Insulator Transition[1]

Y. Imry[2]

IBM Thomas J. Watson Research Center, Yorktown Heights, NY 10598, USA and

Y. Gefen and D.J. Bergman

Department of Physics and Astronomy, Tel Aviv University
Ramat Aviv, Israel

The linear response expressions for the dielectric constant and conductivity as functions of wave number and frequency are evaluated near the localization metal-insulator transition for noninteracting quasiparticles. New behavior is found in the microscopic range. The static dielectric constant from the insulating side is found to diverge with twice the exponent characterizing the vanishing of the conductivity from the metallic side. The optical properties of the system are analyzed in the macroscopic conducting and insulating phases as well as in the microscopic regime. Interesting changes in the behavior for a given material as a function of frequency were found. The dynamic structure factor follows easily from our expressions and we suggest that it can be measured in principle using high-frequency electromagnetic radiation.

1. Introduction

The ability of a metal to screen the Coulomb interaction totally at large distances is an important characteristic of the conducting state of matter. This is related to the wave-vector ($\mathbf{q}$) dependent static dielectric constant[1], $\varepsilon(\mathbf{q},\omega = 0)$ diverging as q^{-2} when $q \rightarrow 0$. In the insulating phase, however, $\lim_{q\rightarrow 0} \varepsilon(\mathbf{q},0) \equiv \varepsilon_0$ is finite. Near a continuous metal-insulator transition, one would thus expect a diverging ε_0 from the insulating side. This is shown here to be the case for the localization, disorder-induced transition, according to its scaling theory[2]. Another important characteristic of the metallic state are its optical properties; e.g., a strong reflectance at all frequencies up to the visible or ultraviolet. This should also change near the metal-insulator transition, with various cross-over effects in the behavior as a function of frequency. Interesting and unusual frequency dependence of both $\varepsilon(\mathbf{q},\omega)$ and conductivity $\sigma(\mathbf{q},\omega)$ may be expected around the metal-insulator transition, resulting in rather exotic optical properties.

---

[1]Research supported in part by the U.S. - Israel Binational Science Foundation (BSF), Jerusalem, Israel.

[2]On leave of absence from the Department of Physics and Astronomy, Tel Aviv University, Ramat Aviv, Israel

In fact, such anomalies are known to exist[3] near the percolation threshold in a metal-insulator macroscopic composite.

In this paper we establish the critical behavior of $\varepsilon(\mathbf{q},\omega)$ and $\sigma(\mathbf{q},\omega)$ near the localization transition for a degenerate fermion system in a self-consistent potential for frequencies much smaller than the inverse elastic relaxation time. Important differences exist between this case and that of the macroscopic composite. These may allow an experimental discrimination between a genuine localization transition and the one dominated by large scale inhomogeneities, which result in percolation-type behavior. We also present results for $\sigma(\mathbf{q},\omega)$ and $\varepsilon(\mathbf{q},\omega)$ in the "microscopic" range, relevant for higher frequencies (or temperatures) and smaller wave lengths, where different power laws are obtained. From the theoretical point of view, the understanding of the correlation function[4] related to the dielectric response[5] may help in elucidating the physical meaning of the order parameter associated with localization. Understanding the screening is also crucial in order to incorporate properly the electron-electron Coulomb interaction effects,[6] of relevance in real physical systems. In fact, similar phenomena may well occur near the Mott-Hubbard metal-insulator transition caused by the interactions[7] even without disorder. Interesting speculations related to this have been made by Toulouse.[8] Recent experiments on doped Si[9] are in a good agreement with the scaling laws for the static properties obtained here. We shall consider in this paper only the three-dimensional case where a mobility edge is believed to exist.

A simple scaling-type argument[10] giving heuristically the scaling law relating the diverging ε_0 from the insulating side to the vanishing $\sigma(0,0)$ from the metallic side follows by considering Fig.1.

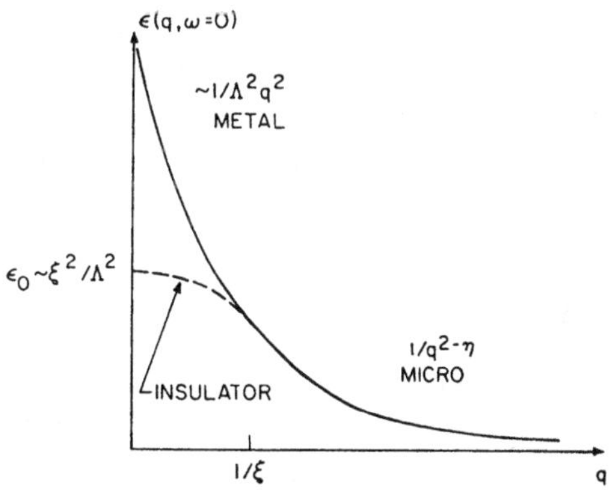

Fig. 1 The static dielectric constant, $\varepsilon(q,0)$, as a function of a q (schematic). For $q>>1/\xi$, both the metal and the insulator are expected to have a dielectric constant $\sim A/q^{2-\eta}$ where A is a constant. For $q<<1/\xi$ the metal should show screening $(\varepsilon(q))\sim q^{-2})$ while for the insulator, $\lim_{q\to 0} \varepsilon(q,0)$ is a finite constant, ε_0

Here, the static dielectric constant $\varepsilon(\mathbf{q},0)$ is schematically depicted as a function of q for the metallic and insulating phases in both the "microscopic" $(q>>1/\xi)$ and the macroscopic $(q<<1/\xi)$ ranges[11]. In the former range the two phases are indistinguishable, displaying in general a $q^{-2+\eta}$ behavior of ε, where η is the usual critical exponent associated with the anomalous dimension of ε. In the latter range the metallic phase is expected to display screening, $\varepsilon(\mathbf{q},0)\cong 1/\Lambda^2 q^2$, where Λ is the screening length, while the insulating phase is expected to have an $\varepsilon(\mathbf{q},0)$ tending to a finite limit, ε_0, as $q \rightarrow 0$. Since as $q\rightarrow 1/\xi$ from below the dielectric functions for both phases should be similar, and because $\varepsilon(\mathbf{q},0)$ does not change much for $q<1/\xi$ in the insulator, it follows that the order of magnitude of ε_0 should be

$$\varepsilon_0 \sim \xi^2/\Lambda^2 \sim \xi^{2-\eta}, \tag{1}$$

(one has $\Lambda^2 \sim \xi^{\eta}$ by matching the above two forms at $q\sim 1/\xi$) i.e., it diverges with $(2-\eta)$ times the exponent characterizing the vanishing[2,12] of $\sigma(0,0)$ (given by $\sim e^2/\hbar\xi$) when the transition is approached from the metallic phase. This relation, with $\eta\cong 0$, is consistent with the experimental results of ref. 9. The same exponent ratio becomes on the order of .3-.5 for the macroscopic composite case[3]. One may feel that this difference in the ratio is large enough to make the two cases experimentally distinguishable. The approximate calculations reported here for noninteracting quasiparticles support the picture of Fig.1 and (1) and turn out to produce a value of $\eta = 0$ so that the above exponent ratio is exactly *two*. The total frequency-dependent $\varepsilon(\mathbf{q},\omega)$ and $\sigma(\mathbf{q},\omega)$ will be given and the ensuing optical properties analyzed. Our expressions enable one to easily obtain the dynamic structure factor $S(\mathbf{q},\omega)$ for the disordered system - a quantity which is in principle measurable.

Recently, MCMILLAN[13b] constructed a scaling theory[2,12a] with Coulomb interaction effects. Near the fixed points *with interactions*, he finds (1) with $0<\eta<2$ (his η is 3 minus ours). Our results are near the *no interaction* fixed point and can be obtained from [13b] by putting $\eta=0$ (or MCMILLAN'S η to be 3). However, we obtain full screening throughout the metallic phase and at the transition. We believe that our treatment clarifies the relations between diffusion and screening in the various regimes.

2. General Results for $\sigma(\mathbf{q},\omega)$ and $\varepsilon(\mathbf{q}\ \omega)$ in the various regimes

We start from the linear response formula[1] for the complex dielectric function $\tilde{\varepsilon}(\mathbf{q},\omega) = \varepsilon(\mathbf{q},\omega) + \frac{4\pi i}{\omega}\sigma(\mathbf{q},\omega)$ which is exact for noninteracting fermions

$$\tilde{\varepsilon}(\mathbf{q},\omega) = 1 - \lim_{\delta\rightarrow 0} \frac{4\pi e^2}{Vq^2}\int dE_i N(E_i)\int dE_f N(E_f)\,|\langle f|e^{i\mathbf{q}\cdot\mathbf{r}}|i\rangle|^2 \times \frac{f(E_i)-f(E_f)}{E_i-E_f+\hbar\omega+i\delta} \tag{2}$$

where V is the volume of the system, N(E) the density of states per unit energy whose variation with E for small E will be neglected (later we denote $n(E) \equiv N(E)/V$), f(E) is the Fermi function, E is measured from the Fermi energy and $|i\rangle$ and $|f\rangle$ are exact single-particle states. Following suggestions in [6c] and [13] one can derive a relation between the matrix elements in (2) and a dynamic structure factor[14] $S_o(\mathbf{q},\omega)$, for quantum diffusion for a single particle defined in the following way. At t=0 the particle is started with a wave packet around the origin which is much narrower than ξ in space and is also of minimal width in energy. Then

$$S_o(\mathbf{q},\omega) = \sum_{i,f} p_i \, |\langle f| e^{i\mathbf{q}\cdot\mathbf{r}} |i\rangle|^2 \delta(E_f - E_i - \hbar\omega) = \int dt \, e^{i\omega t} F(\mathbf{q},t) \tag{3}$$

where, for a gaussian distribution (in a semi-classical approximation), which should hold after many elastic collisions

$$F(q,t) = \exp\left[-\frac{1}{2} q^2 \langle r(t)^2 \rangle\right] \tag{4}$$

and p_i are the weights of the initial states $|i\rangle$. Assuming that for the narrow range of E_i considered, the matrix elements squared depend only on energy differences, we find

$$|\langle f| e^{i\mathbf{q}\cdot\mathbf{r}} |i\rangle|^2 = S_o(q,\omega)/N(0), \quad \omega = E_f - E_i \tag{5}$$

and $\hbar$ is taken as unity. Using (2) and (5) we immediately find

$$\sigma(\mathbf{q},\omega) = \frac{\pi\omega^2 e^2}{q^2} n(0) S_o(\mathbf{q},\omega) \quad \left[\underset{q \ll 1/\xi}{\rightarrow} \frac{e^2 n(0) D\omega^2}{(Dq^2)^2 + \omega^2} \quad \text{or} \quad \frac{\sigma(0,0)\omega^2}{(Dq^2)^2 + \omega^2}, \right] \tag{6}$$

where to get the last expression we used the Einstein relation between D and σ for degenerate Fermions and the linear increase $\langle r^2 \rangle = 2Dt$ for macroscopic times. From this it follows[14b] by (4) that $S_o(\mathbf{q},\omega) = Dq^2/\pi[(Dq^2)^2 + \omega^2]$. Here D is the macroscopic diffusion constant which, within the scaling theory,[2] vanishes like ξ^{-1} when localization is approached from the conducting side. For $q \ll 1/\xi$ the macroscopic expressions are sufficient (see Figs.2a,b).

This is valid for the conducting phase, where the macroscopic D is finite. It is now straightforward to obtain the real part corresponding to (6) of the dielectric constant.

$$\varepsilon(\mathbf{q},\omega) = 1 + \frac{4\pi n(0) e^2 D \cdot Dq^2}{(Dq^2)^2 + \omega^2}. \tag{7}$$

These expressions[6c], valid at low frequencies, yield the optical properties of the conductor for $q \ll 1/\xi$. We would now like to extend them to the microscopic ($q \gg 1/\xi$) range and to the insulating phase.

Let us start with $\omega = 0$ to simplify the calculation, and take $Dq^2 \ll kT \ll E_F$ (assuming a degenerate Fermi system). Essentially similar results hold also at the

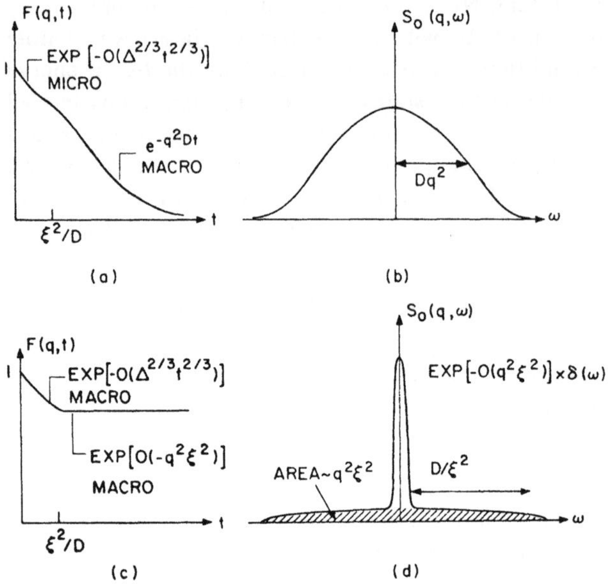

Fig. 2 Schematics of F(q,t) and $S_o(q,\omega)$ in the metallic and insulating phases for $q<<1/\xi$.
a) F(q,t) for the metal, changes around $t \sim \xi^2/D$ from $\exp[-O(\Delta^{2/3}t^{2/3})]$ to $\exp(-q^2Dt)$.
b) $S_o(q,\omega)$ for the metal, a Lorentzian of width Dq^2.
c) F(q,t) for the insulator, saturates at $e^{-O(q^2\xi^2)}$ for $t > \xi^2/D$.
d) $S_o(q,\omega)$ for the insulator, consists of a dominant $\delta(\omega)$ part and a broad background of weight $O(q^2\xi^2)$. Here D is the macroscopic diffusion constant for the metal at the same ξ.

$T \to 0$ limit. The Fermi function is now smooth on the relevant range of $(E_f - E_i)$ and one may replace $[f(E_f)-f(E_i)]/(E_f-E_i)$ by $f'(E_i)$ to get

$$\varepsilon(\mathbf{q},0) = 1 + \frac{4\pi e^2}{q^2} n(0) P \int_{-\infty}^{\infty} S_o(\mathbf{q},\omega) d\omega. \tag{8}$$

$S_o(\mathbf{q},\omega)$ satisfies the sum rule $\int S_o(\mathbf{q},\omega)d\omega = F(\mathbf{q},0) = 1$. Thus $\varepsilon(\mathbf{q},0) = 1 + 1/\Lambda^2 q^2$ where $\Lambda = [4\pi e^2 n(0)]^{-1/2}$ is the Thomas-Fermi screening length. (Notice that in this picture we ignore the electron-electron interactions. Inclusion of these interactions[13b] induces a change in the density of states and the screening length may thus be different.) This is valid in the whole metallic range, including the macroscopic and microscopic domains (i.e. $\eta = 0$). The situation in the insulating phase is different, however. The wave packet at T=0 consists only of states that are localized near the origin and thus can never escape from a volume of the order of ξ^3, so that as $t\to\infty$, $r^2(t)\to O(\xi^2)$. From this and the microscopic range, discussed later, follow the schematic forms of F $(\mathbf{q},t)$ and $S_o(\mathbf{q},\omega)$ for the insulator depicted in Figs.2c and 2d. The localization thus endows $S_o(\mathbf{q},\omega)$ with a $\delta(\omega)$ part which is dominant for $q<<1/\xi$. It does not contribute to the principal part of (8) and only the small smooth background of area $O(\mathbf{q}^2\xi^2)$ contributes. Thus, for the insulating phase, for $q<<1/\xi$, $\varepsilon(\mathbf{q},0)\sim\xi^2/\Lambda^2$ in agreement with the heuristic expectation of eq. (1) with $\eta=0$. The background part (Fig.2d) of $S_o(\mathbf{q},\omega)$ was also estimated explicitly. Using (6) we obtain an expression for the conductivity in the insulating phase for $q<<1/\xi$. We interpolate between the low $(\omega<<D/\xi^2)$ and high $(\omega>>D/\xi^2$, similar to (12) below) frequency regimes,

$$\sigma(\mathbf{q},\omega)\sim\frac{\omega^2 e^2 n(0)(D\xi)^{2/3}}{A_1\omega^{5/3}+A_2(D/\xi^2)^{5/3}}, \tag{9}$$

where here and in the following expressions A_i $(i=1,2,...)$ stand for numerical coefficients of order unity. At low frequencies the leading term is $\sigma(\mathbf{q},\omega)\sim\sigma_m(0,0)[\omega/(D/\xi^2)]^2$, where $\sigma_m(0,0)$ is the metallic conductivity at the same ξ $(\sim e^2/\hbar\xi)$. This divergence of the coefficient of ω^2 in $\sigma(\omega)$ as the transition is approached was first obtained by WEGNER[12b]. The dielectric constant in the insulating regime is likewise given by an interpolation formula (where ω signifies an absolute value of ω)

$$\varepsilon(q,\omega)=1+4\pi e^2 n(0)\frac{(D\xi)^{2/3}}{A_3\omega^{2/3}+A_4(D/\xi^2)^{2/3}}. \tag{10}$$

We now focus our attention on the microscopic behavior for $t<\xi^2/D$, which dominates what happens at $q>1/\xi$ and/or $\omega>D/\xi^2$. According to the scaling theory[2,16] the diffusion constant D(L) is length-scale (L) dependent, and in fact it is proportional in three dimensions to $1/L$ for $L<<\xi$. It thus follows[16] that $<r^2(t)>\sim t^{2/3}$, and $F(q,t)=e^{-(\Delta t)^{2/3}}$ where $\Delta\sim D\xi q^3$, D again being the macroscopic diffusion constant. It is now possible only to evaluate $S_o(\mathbf{q},\omega)$ for $\omega<<\Delta$ and $\omega>>\Delta$, a convenient interpolation formula being

$$S_o(q,\omega)=\frac{A_5\Delta^{2/3}}{A_6\Delta^{5/3}+\omega^{5/3}}. \tag{11}$$

This yields from (6) the conductivity in the microscopic (both metal and insulator!) range.[17]

$$\sigma(q,\omega)=\frac{A_5\pi\omega^2 n(0)e^2(D\xi)^{2/3}}{\omega^{5/3}+A_6\Delta^{5/3}}. \tag{12}$$

For $\omega\sim\tau_{in}^{-1}$ and $q\sim\ell_{in}^{-1}$ where $\ell_{in}^3\sim D\xi\tau_{in}$, this agrees with the dc conductivity in the microscopic range previously discussed[16], and has important consequences as far as the temperature dependence of the low-frequency conductivity is concerned. One can also evaluate $\varepsilon(\mathbf{q},\omega)$ in this range, which again yields an interpolation form:

$$\varepsilon(\mathbf{q},\omega) = 1 + \frac{4\pi e^2 n(0)}{q^2} \frac{A_7\Delta^{2/3}}{A_8\omega^{2/3} + \Delta^{2/3}}. \tag{13}$$

The static $\varepsilon(\mathbf{q},0)$ can be evaluated at T=0, which again yields $\varepsilon(\mathbf{q},0) = 1 + 1/q^2\Lambda^2$. which is valid for noninteracting Fermions for $q>>1/\xi$ in both the metallic and the insulating phases. In the former this is valid, of course, also for $q<<1/\xi$

3. Analysis of the Optical Properties

We employ the expressions derived for the longitudinal ε and σ in the various regimes to calculate the electromagnetic properties of the system. This procedure is valid for electromagnetic waves as long as the longitudinal and transverse responses of the system are similar.

We just need the complex refractive index given by

$$n + ik = [\varepsilon(q,\omega) + 4\pi i\sigma(q,\omega)/\omega]^{1/2} \tag{14}$$

n and k determine all the optical properties of the material. We shall display results for the absorption coefficient[19]

$$\eta = 2k\omega/c \tag{15}$$

where c is the velocity of light in vacuum and the reflection coefficient for normal incidence:

$$R = \frac{(n-1)^2 + k^2}{(n+1)^2 + k^2}. \tag{16}$$

It is important to note that n and k are really functions of ω or of q only, since q and ω for electromagnetic waves are related by the dispersion relation.

$$\omega(q) = (c/n)q. \tag{17}$$

It can be shown that, due to the high value of c, one can self-consistently neglect the q-dependent terms in the denominators of (6), (7), (11-13).

Before analyzing the optical properties, we establish some order of magnitude relations. It is useful to remember that the Fermi energy and the typical (unscreened) Coulomb energy $E_C \equiv e^2 k_F$ are of the same order of magnitude in most metallic systems. To cover cases such as extremely low electron density semiconductor systems in their metallic phases (both in the bulk and in 2D inversion layers), we shall keep the ratio E_C/E_F in our equations. Due to the extremely weak electron density dependence of this ratio, it probably cannot change, however, by more than an order of magnitude for all metallic systems. It is interesting to note that the ratio of the Fermi velocity v_F to c is on the order of

$$v_F/c \sim \alpha \frac{E_F}{E_C} \sim 10^{-2} \tag{18}$$

where α is the fine-structure constant $\alpha \equiv e^2/\hbar c \cong 1/137$. We note also that the product of the Fermi wave number and the Thomas-Fermi screening length is of the order of magnitude of

$$k_F \Lambda \sim (E_F/E_C)^{1/2}. \tag{19}$$

We are interested in the range where the disorder is large enough for localization to be relevant, i.e., $k_F \ell \gtrsim 1$, $E_F \tau \gtrsim 1$, $k_F \xi \gtrsim 1$ (or $k_F \xi >> 1$, near the transition). We also discuss frequencies that are low enough so that the scattering is important, i. e.,

$$\omega\tau \lesssim 1. \tag{20}$$

We remember that the Plasma frequency, ω_p, is given by $\hbar\omega_p \sim (E_F E_C)^{1/2}$, so that $\omega_p \tau \gtrsim 0(1)$. For $\omega\tau >> 1$, one may expect the scattering to be relatively unimportant. Finally, we observe that since the metallic conductivity is, in our range, on the order of [2,12] $e^2/\hbar\xi$, the diffusion constant is given through the Einstein relation by

$$D \sim [n(0)\xi]^{-1}. \tag{21}$$

The ratio $c\xi/D$ is

$$c\xi/D \sim (c/v_F)(k_F \xi)^2 >> 1. \tag{22}$$

This means physically that the time for light to propogate a distance ξ is much smaller than the time it takes to the electron to diffuse the same length (i.e. $\xi/c << \xi^2/D$).

We can distinguish three regimes of distinct optical behavior:

a. Macroscopic Metallic phase, $q\xi << 1$ and $\omega << D/\xi^2$.

Here (6) holds for the conductivity which is the dominant part in the complex dielectric function (see (14)). We find the usual metallic results (σ is the dc conductivity)

$$n \sim k \sim (\sigma/\omega)^{1/2} \tag{23}$$

$$k = 1 - 0(\sqrt{\omega/\sigma}) \tag{24}$$

$$\eta \sim \sqrt{\sigma\omega}/c. \tag{25}$$

It turns out that to get the macroscopic range (7) for the dielectric constant one needs the stronger condition $\omega << Dq/\xi$. For $\omega >> Dq/\xi$, the microscopic expression (13) holds. This is relevant for corrections to the principal behavior given by (23-25).

b. Macroscopic Insulating Phase, $q\xi << 1$ and $\omega << D/\xi^2$.

Here, Eqs. (9) and (10) hold and we find near the transition a dielectric with a high index of refraction:

$$n \sim \xi/\Lambda, \; k \sim (\xi/\Lambda)\omega/(D/\xi^2), \; k << n \tag{26}$$

$$R = 1 - 0(\frac{\Lambda}{\xi}), \tag{27}$$

where the interesting dominant frequency-dependent correction to R comes not from k but from the weak frequency dependence of ε alluded to in section 2, and:

$$\eta \sim \frac{\xi}{\Lambda} \frac{\omega^2}{cD/\xi^2}. \tag{28}$$

c. Microscopic Regime $q\xi >> 1$ or $\omega >> D/\xi^2$ (in *both* the metal and the insulator)

To get to $q\xi >> 1$ one needs to be very close to the metal-insulator transition, thus this regime will usually be obtained at higher frequencies only. Here we have[17] a behavior characterized by typical[12,17] power laws:

$$n \sim k \sim (\frac{E_C}{E_F})^{1/2} (\frac{E_F}{\hbar\omega})^{1/3} \tag{29}$$

$$R = 1 - 0\left[\sqrt{E_F/E_C}\,(\hbar\omega/E_F)^{1/3}\right] \tag{30}$$

$$\eta \sim \sqrt{E_c/E_F}\,\frac{\omega^{2/3}(E_F/\hbar)^{1/3}}{c}. \tag{31}$$

Thus, in the metallic phase the system behaves ordinarily at low frequencies, and crosses over to the "microscopic" behavior, characterized by different power laws (29-31), around $\omega \sim D/\xi^2$. In the insulating phase the system crosses over from a large n dielectric behavior to the microscopic behavior around the same frequency as above. This last case is an example of a system which behaves like a quasi metal at high frequencies and like a dielectric at low frequencies, where the cross-over frequency is tunable, in principle. In a real case one has to make sure that the temperature is low enough to be irrelevant (at finite temperatures one has, for example, to intorduce a low-frequency cut-off given by the inverse inelastic time). We emphasize that our results are for noninteracting fermions.

The interactions should introduce a non zero η and accordingly change the power laws in (29-31). Thus, obtaining these power laws experimentally is an important way to study these systems.

4. The Dynamic Structure Factor for the Electrons

Finally, it is straightforward to evaluate from our expressions and well-known general relationships[1] the physical dynamic structure factor $S(q,\omega)$ for the degenerate disordered Fermi system (which is of course different from the hypothetical $S_0(q,\omega)$ used here). We employ the relationship:

$$S(q,\omega) = \frac{q^2}{2\pi e^2} \frac{4\pi\sigma(q,\omega)/\omega}{\varepsilon^2(q,\omega) + (4\pi\sigma(q,\omega)/\omega)^2} \qquad (32)$$

with the σ and ε appropriate to the various regimes discussed above. We would like to suggest the $S(q,\omega)$ can be measured in principle by scattering high-frequency electromagnetic radiation to which the system is relatively transparent. For low-density conductors such as In_2O_{3-x}, this includes visible light[20]. This may prove to be a further useful way to probe the details of these systems, by varying in principle both the length scale and time scale (related to q and ω respectively).

5. Summary

To summarize: The linear response expressions for $\varepsilon(\mathbf{q},\omega)$ and $\sigma(\mathbf{q},\omega)$ can be evaluated around the localization transition. In particular, the static dielectric constant for the insulator is shown to diverge, for noninteracting quasiparticles, with an exponent twice that characterizing the vanishing of the conductivity from the metallic side, a result which is quite different from the macroscopic composite case. New expressions and power-laws are found in the microscopic range. All these quantities appear to satisfy a sort of dynamic scaling[5a] which will be discussed elsewhere. The optical properties in the various regimes were analyzed and the dynamic structure factor for the electrons obtained. It is suggested that the latter can be measured and yield new information on these systems.

Acknowledgements

We thank M. AZBEL, E. FRADKIN, A. HOUGHTON and Z. OVADYAHU for useful conversations. We are grateful to F. WEGNER for pointing out to us during a discussion at this Symposium an error in the initial manuscript of [18]. The hospitality of the Physics Department at the Brookhaven National Laboratory where this work was started is gratefully acknowledged by Yoseph Imry and Yuval Gefen. Yoseph Imry is likewise grateful to the Institute for Theoretical Physics, University of California Santa Barbara, where this work was concluded.

References

[1] See e.g. P. Nozieres, *Interacting Fermi Systems*, Benjamin (1964); C. Kittel, *Quantum Theory of Solids*, Wiley (1963).

[2] E. Abrahams, P. W. Anderson, D. C. Licciardello and T. V. Ramakrishnan, Phys. Rev. Lett. **42**, 673 (1979).

[3] a) A. L. Efros and B. I. Shklovskii, Phys. Status Solidi B**76**; 475 (1976);
b) V. E. Dubrov, M. E. Levinshtein and M. S. Shur, Z. Eksper. Teor. Fiz. **70**, 2014 (1976), [Sov. Phys. JETP **44**, 1050 (1976)];
c) D. J. Bergman and Y. Imry, Phys. Rev. Lett. **39**, 1222 (1977);
d) This type of effect was seen experimentally by D. M. Grannan, J. C. Garland and D. B. Tanner, Phys. Rev. Lett. **46**, 375,(1981).

[4] a) A. Aharony and Y. Imry, J. Phys C**10**, L487 (1977);
b) H. G. Schuster, Z. Phys. **31**, 99 (1978).

[5] a) W. Götze, J. Phys. C**12**, 1279 (1979), made an approximate calculation of $\varepsilon(q,\omega)$ and $\sigma(q,\omega)$ near the localization yielding (1) but differing from (9) due to a gap in the excitation spectrum in the insulating phase;
b) D. Vollhardt and P. Wölfle, Phys. Rev. B**22**, 4678 (1980), derived a systematic expansion for the response functions at and below two dimensions. P Wölfle (these proceedings) discussed the generalization of this work (D. Vollhardt and P. Wölfle, unpublished) to higher dimensions.

[6] a) B. L. Altshuler and A. G. Aronov, Zh. Eksper. Teor. Fiz. **77**, 2028 (1979) [Sov. Phys. JETP **55**, 968 (1979)].
b) B. L. Altshuler, A. G. Aronov and P. A. Lee, Phys. Rev. Lett. **44**, 1288 (1980);
c) P. A. Lee, unpublished.

[7] N. F. Mott, *Metal Insulator Transitions*, (Taylor and Francis, London, 1974).

[8] G. Toulouse, J. Phys. **36**, 1137 (1975).

[9] a) M. Capizzi, G. A. Thomas, F. de Rosa, R. N. Bhatt and T. M. Rice, Solid State Commun. **33**, 793 (1980).
b) T. F. Rosenbaum, K. Andres, G. A. Thomas and R. N. Bhatt, Phys. Rev. Lett. **45**, 1723 (1980).
c) M. Capizzi, G. A. Thomas, F. de Rosa, R. N. Bhatt and T. M. Rice, Phys. Rev. Lett. **44**, 1019 (1980).
d) T. G. Castner, N. K. Lee, G. S. Cieloszyk and G. L. Salinger, Phys. Rev. Lett. **34**, 1627 (1975).

10 A similar heuristic argument had been previously given in [9b]. The authors are grateful to G. A. Thomas and F. Wegner for useful discussions on this subject. In the present work, the independence of Λ on disorder for noninteracting quasiparticles is established.

11 ξ is the characterisitic coherence length around the localization transition[2]. Its physical significance in the metallic state is discussed by Y. Imry, Phys. Rev. Lett **44**, 469 (1980).

12 a) F. Wegner, Z. Phys. **25**, 327 (1976).
b) F. Wegner, pp. 302 in: Amorphous and Liquid Semiconductors, Proc. of the 7th International Conference, W. E. Spear, ed., University of Edinburgh (1977).

13 a) E. Abrahams, P. W. Anderson, P. A. Lee and T. V. Ramakrishnan, unpublished.
b) W. L. McMillan, Phys. Rev. B**24**, 2739 (1981).
c) M. Kaveh and N. F. Mott, J. Phys. C**14**, L177, (1981).
d) M. Azbel has remarked to us that these expressions also follow easily from the well-known, quasi-classical approximation for the matrix elements.

14 a) L. Van Hove, Phys. Rev. **95**, 249 (1954);
b) G. Vineyard, Phys. Rev. **110**, 99 (1958).

15 Apart from logarithmic corrections, the low frequency behavior of (9) agrees in order of magnitude with Mott's result in the insulating phase (N. F. Mott and E. A. Davis, *Electronic Processes in Non-Crystalline Materials* (Pergamon, 2nd edition, Oxford, 1979) for strong localization, where ξ becomes comparable to the microscopic length. However, (9) predicts a ξ^5 increase[12b] of the coefficient of ω^2 when the insulator becomes weak, approaching the transition with diverging ξ, along with a vanishing of the range of the ω^2 behavior.

16 Y. Imry, J. App. Phys. **52**, 1817 (1981).

17 Eqs. (9) and (10) imply that at the metal-insulator transition $\sigma(0,\omega)\sim\omega^{1/3}$, a result[12] which has recently also been obtained by B. Shapiro and E. Abrahams, unpublished. From our expression for ε in the insulating phase and (11), we find likewise that at the transition $\varepsilon(0,\omega)\sim\omega^{-2/3}$. These results were first reported in [18].

18 Y. Imry, Y. Gefen and D. J. Bergman to be published

19 J. Ziman, Principles of the Theory of Solids, Cambridge (1969).

20 Z. Ovadyahu and Y. Imry, Phys. Rev. B, Rapid Communications, in Press; Z. Ovadyahu, private communication.

Localization in *n*-InSb

K. Hoshino

Faculty of Engineering, Niigata University
Niigata 950-21, Japan

It is shown that the characteristic features of the carrier density, the magnetic-field and the temperature dependences of the magnetoresistance in metallic n-InSb observed recently by Morita et al. can be explained qualitatively by a theory, in which the effect of the Coulomb interactions as well as the localization effect are taken into account. The competition between these effects is emphasized.

1. Introduction

Recent development in the theory of the Anderson localization enables us to discuss the temperature and the magnetic-field dependences of the resistivity of disordered systems. So far much theoretical and experimental work on two-dimensional disordered systems such as Si-MOS and metallic films has been reported. On the other hand, there has been little work on three-dimensional systems [1]. Quite recently, however, very interesting experimental results have been reported for n-InSb [2] and for Si:P [3], where quantum correction due to the localization effect and the effect of the Coulomb interaction on the conductivity is expected to be observed. As InSb is characterized by the very small isotropic effective mass ($m = 0.014m_0$, m_0 being free electron mass) and the single valley, this system is simpler than other doped semiconductors such as Si:P and Ge:Sb, where effects of the valleys and the anisotropy of mass will play important roles. Therefore, we consider that n-InSb is suitable to check a theory.

The purpose of this paper is to explain the experimental results of the magnetoresistance (MR) in n-InSb by our theory, in which the effect of the Coulomb interaction as well as the localization effect are taken into account. First, we summarize briefly the characteristic features of the experimental results in section 2 and then show how we can explain these features theoretically in section 3.

2. Experimental Results

Morita et al.[2] have reported remarkable experimental data on the temperature and the magnetic-field dependences of the resistivity of n-InSb down to 20 mK. Their samples are metallic and the carrier density ranges from 1.2 to 9.3×10^{14} cm^{-3}.

The temperature dependences of the resistivity at H=0 and H=100 Oe are shown in Figs.1 and 2 for two choices of the carrier density. When carrier density is high ($n = 6.3\times10^{14}$ cm^{-3}), MR is always negative and its absolute magnitude increases with increasing temperature. On the other hand, when carrier density is low ($n = 1.6\times10^{14}$ cm^{-3}), MR is positive at low temperature and negative at higher temperatures. Therefore, the temperature dependence of MR changes qualitatively with decreasing carrier density.

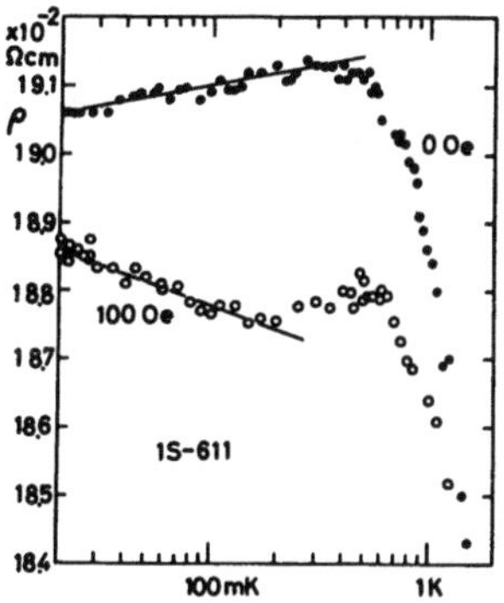

Fig.1 Temperature dependence of the resistivity ($n = 6.3\times10^{14}$ cm^{-3})

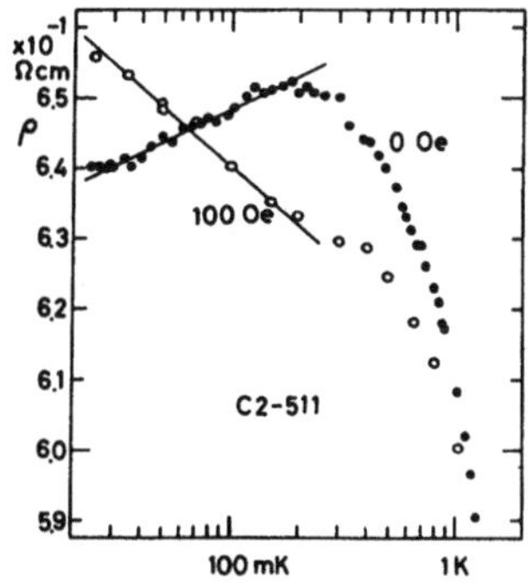

Fig.2 Temperature dependence of the resistivity ($n = 1.6\times10^{14}$ cm^{-3})

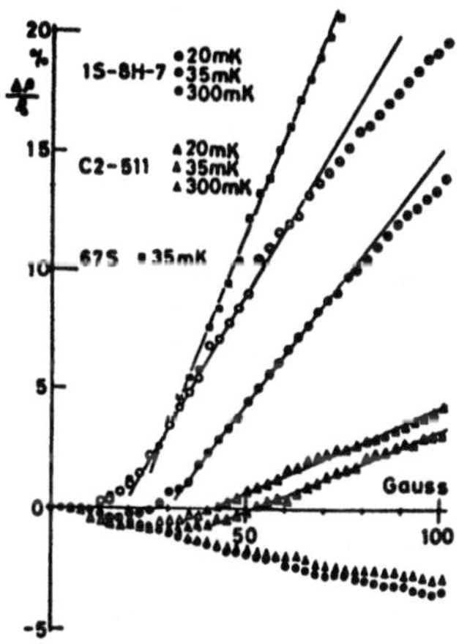

Fig.3 Magnetic-field dependences of the magnetoresistance

It is very interesting to see the magnetic-field dependence of MR. Figure 3 shows MR of samples with rather low carrier density as a function of the magnetic field for various temperatures. The characteristic feature is as follows; even when MR is mainly positive, it is negative at very weak magnetic field at low temperature, whereas MR is always negative at high temperatures.

In the next section, we show how we can explain these experimental features theoretically.

3. Theory

Recent perturbation theory in the weakly localized regime [4] has been applied to three-dimensional disordered systems by KAWABATA [1]. He has shown that the localization theory gives always negative MR. On the other hand, as was shown in the previous section, MR of n-InSb is either positive or negative. Therefore, we have to take into account other mechanisms, which give the positive contribution to MR.

One possibility is the effect of the spin-orbit interaction. It was shown by HIKAMI et al.[5] that the effect gives rise to the positive contribution to MR. MAEKAWA and FUKUYAMA [6] explained successfully the experimental results for MR of Cu metallic films by considering the competition between the effects of localization and the spin-orbit interaction, where they also took into account the effect of the Zeeman spin-splitting. In a similar way, FUKUYAMA and HOSHINO [7] have calculated MR of n-InSb by including the effects of localization, spin-orbit interaction and Zeeman spin-splitting and shown that the overall behaviour, namely MR is positive at low temperatures and negative at higher temperatures, is consistent with the experimental results. However, it is difficult to explain by this theory the following two features; (i) even when MR is mainly positive, it is negative at very weak magnetic field and (ii) the carrier density dependence of the sign of MR. Thus, this theory can not be considered as satisfactory.

The other possibility is the effect of Coulomb interaction, which has been studied actively quite recently. In the absence of magnetic field, quantum correction to the conductivity of disordered systems due to Coulomb interaction has been studied by FUKUYAMA [8] and ALTSHULER et al.[9] independently. At finite magnetic field, the effect of Coulomb interaction on MR has been investigated by FUKUYAMA [10] and ALTSHULER et al.[11] for orbital effect and by LEE and RAMAKRISHNAN [12] and by KAWABATA [13] for Zeeman effect. In the following, we apply FUKUYAMA's theory to n-InSb.

According to FUKUYAMA [10], we classify the scattering processes due to Coulomb interaction together with the particle-particle and the particle-hole impurity scattering ladder by using g_i's (i= 1-4). The contributions due to these scattering processes to MR are given, in general, as follows;

g_1 process --- no contribution
g_2 process --- orbital effect
g_3 process --- Zeeman effect
g_4 process --- <u>both</u> orbital and Zeeman effects.

The magnetoconductivity $\Delta\sigma_{Coul}(H,T)$ due to Coulomb interaction can be written as

$$\begin{aligned}&\Delta\sigma_{Coul}(H,T)/(\sigma_0 \frac{3\sqrt{3}}{8} \frac{1}{\varepsilon_F^2\tau^2})\\ &\equiv \Delta\tilde{\sigma}_{Coul}(H,T)\\ &= g_2\phi_{orb}(H) - g_3\phi_{Zeeman}(H) - g_4(\phi_{orb}(H) + \phi_{orb\text{-}Z}(H)),\end{aligned} \tag{1}$$

where $\sigma_0 = ne^2\tau/m^*$ and $\phi_{orb}(H)$, $\phi_{Zeeman}(H)$ and $\phi_{orb\text{-}Z}(H)$ are functions of magnetic field.

In the following, we consider orbital effect only. In this case, $\phi_{Zeeman}(H)=0$ and $\phi_{orb-Z}(H) = \phi_{orb}(H)$ and then (1) reduces to

$$\Delta\tilde{\sigma}_{Coul}(H) = (g_2 - 2g_4)\phi_{orb}(H). \quad (2)$$

The reason why we consider only orbital effect is as follows: The characteristic magnetic fields of orbital and Zeeman effects are given by

$$H_{orb} = \frac{8}{3}\varepsilon_F\tau\omega_c/\max(T,1/\tau_\varepsilon) = \frac{8}{3}\varepsilon_F\tau\omega_c/T \quad \text{for } T\tau_\varepsilon \gg 1 \quad (3)$$

and

$$H_{Zeeman} = |g^*|\mu_B H/T \quad (4)$$

where ω_c is the cyclotron frequency. The ratio of (3) to (4) is much larger than unity for n-InSb, i.e.

$$H_{orb}/H_{Zeeman} = \frac{8}{3}\varepsilon_F\tau\omega_c/|g^*|\mu_B H = 7.5\varepsilon_F\tau \ . \quad (5)$$

Therefore, the contribution to MR due to orbital effect dominates that due to Zeeman effect in the case of n-InSb ($g^*=-51.3$).

By noting $g_2=g_4=F/2$, F being the quantity introduced by ALTSHULER et al.[9], we finally obtain MR due to Coulomb interaction as

$$-\Delta\tilde{\sigma}_{Coul}(H,T) = (F/2)\phi_{orb}(H). \quad (6)$$

The magnetoresistance due to localization effect and Coulomb interaction is given by

$$\Delta\rho(H,T) = -(\Delta\tilde{\sigma}_{loc}(H,T) + \Delta\tilde{\sigma}_{Coul}(H,T)), \quad (7)$$

where $\Delta\tilde{\sigma}_{loc}(H,T)$ is the magnetoconductivity due to localization effect [1].

In Fig.4 we show $\phi_{orb}(H)$ as functions of magnetic field for $2\pi T\tau = 0.1$, 0.2 and 0.5. For comparison, $\Delta\tilde{\sigma}_{loc}(H,T)$ is also shown in this figure. From (6) and (7), the sign of MR is determined by the competition between the localization (negative) and the Coulomb (positive) contributions. As F decreases with increasing carrier density because of the screening effect, the larger value of F corresponds to the lower electron density.

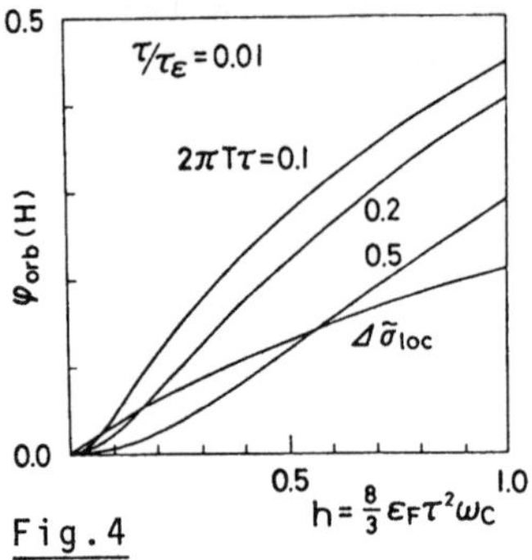

Fig.4

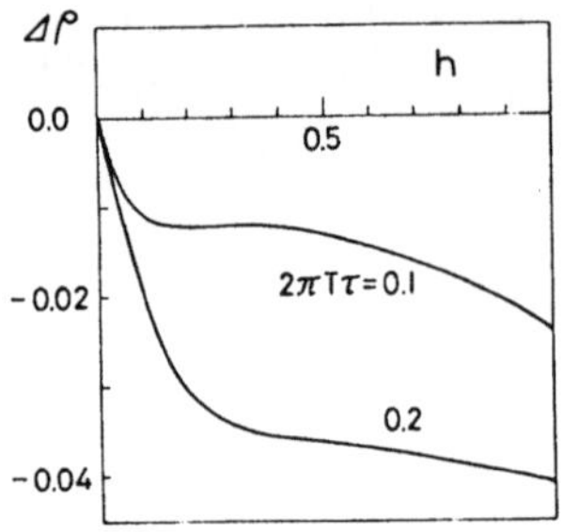

Fig.5

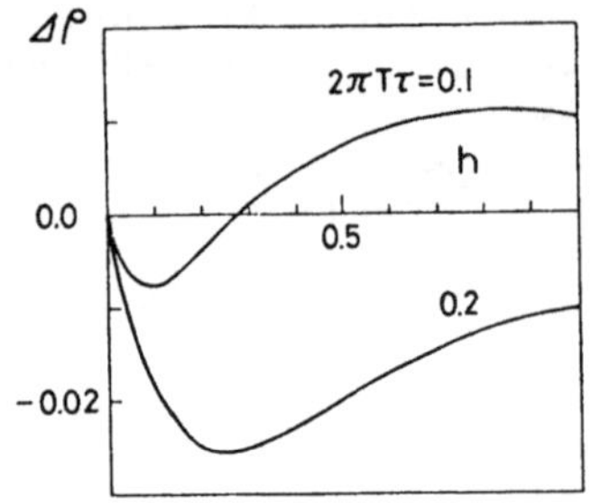

Fig.6

Fig.4 Magnetic field dependence of $\phi_{orb}(H)$

Fig.5 Magnetic field dependence of the magnetoresistance (F=0.85)

Fig.6 Magnetic field dependence of the magnetoresistance (F=1.0)

We show in Figs.5 and 6 the field dependences of MR for small and large F, respectively. When F= 0.85 (higher carrier density), MR is always negative irrespective of magnetic field and its absolute value increases with increasing temperature, which is consistent with the experimental results shown in Fig.1. On the other hand, when F= 1.0 (lower carrier density), MR is negative for weak magnetic field and turns positive for stronger field at low temperatures, though it is always negative at higher temperatures. This result is qualitatively in good agreement with the experimental features shown in Figs.2 and 3.

In conclusion, we can explain qualitatively the temperature and the magnetic-field dependences of the magnetoresistance of n-InSb [2] by taking into account the effect of Coulomb interaction as well as the localization effect.

This report is based on the work studied in collaboration with Professor H.Fukuyama and Dr.Y.Isawa.

References

1. A.Kawabata: Solid State Commun. 34 431 (1980); J.Phys.Soc. Jpn. 49 628 (1980)
2. S.Morita, T.Fukase, Y.Isawa, S.Ishida, Y.Takeuti and N. Mikoshiba: Proc.15th Int.Conf.Physics of Semiconductors, Kyoto, 1980, J.PHYS.Soc.Jpn. 49 Suppl.A, 379 (1980); Proc. 16th Int.Conf.Low Temperature Physics, Los Angels, 1981, Physica 107B 421 (1981); S.Morita: private communication
3. T.F.Rosenbaum, K.Andres, G.A.Thomas and P.A.Lee: Phys. Rev. Lett. 46 568 (1980)
4 L.P.Gor'kov, A.I.Larkin and D.E.Khmelnitskii: J.E.T.P.Lett 30 248 (1979)
5. S.Hikami, A.I.Larkin and Y.Nagaoka: Prog.Theor.Phys. 63 707 (1980)

6. S.Maekawa and H.Fukuyama: J.Phys.Soc.Jpn.50 8 (1981)
7. H.Fukuyama and K.Hoshino: J.Phys.Soc.Jpn.50 2131 (1981)
8. H.Fukuyama: J.Phys.Soc.Jpn. 48 2169 (1980)
9. B.L.Altshuler, A.G.Aronov and P.A.Lee: Phys.Rev.Lett. 44 1288 (1980)
10. H.Fukuyama: J.Phys.Soc.Jpn. 50 3407 (1981)
11. B.L.Altshuler, A.G.Aronov, D.E.Khmelnitskii and A.I.Larkin: preprint
12. P.A.Lee and T.V.Ramakrishnan: preprint
13. A. Kawabata: J.Phys.Soc.Jpn.50 2461 (1981)

Interplay of Disorder and Coulomb Interactions in Anderson-Localized States

H. Kamimura, T. Takemori and A. Kurobe

Department of Physics, Faculty of Science, University of Tokyo
Bunkyo-ku, Tokyo 113, Japan

This paper discusses a general theoretical treatment to investigate the rival effects of disorder and electron-electron interactions in the Anderson-localized regime. First it is shown that the intrastate interaction gives rise to the occurrence of singly occupied states as well as doubly occupied states near and below the Fermi level and that this results in a Curie-type spin susceptibility at low temperatures. Secondly it is shown that electron-electron interactions between Anderson-localized states lead to both ferromagnetic (direct exchange) and antiferromagnetic (kinetic-type exchange) interactions among spins of singly occupied states. Specific heat anomaly beyond the T-linear behavior, in particular its magnetic field dependence, and magnetic susceptibility at very low temperatures are calculated by taking into account the coexistence of ferromagnetic and antiferromagnetic interactions within a spin pair model. It is shown that the experimental results on Si:P can be well explained by the present treatment. Finally the effects of electron-electron interactions on the magnetoresistance of the variable range hopping conduction are investigated. It is shown that magnetoresistance is positive as far as the change of localization length due to a magnetic field is neglected. Taking into account the latter effect which is responsible for negative magnetoresistance, the total magnetoresistance including positive and negative parts is calculated for the system of $1T\text{-}TaS_2$ which is considered to be two dimensional. The peculiar features of magnetoresistance observed in that system can be explained well by the present theory.

1. Introduction

In the intermediate concentration regime of doped semiconductors the electronic states near and below the Fermi level are localized in the Anderson sense and the electron correlation plays an important role in this regime. As pointed out by one of the present authors (H.K.) and his coworkers [1,2,3], the intrastate interaction energy is needed in order for another electron to fill a state already occupied by an electron, and thus at and below the Fermi level there are states which are occupied by only one electron (singly occupied states) as well as those occupied by two electrons (doubly occupied states). Each singly occupied state carries a localized spin of $S=1/2$ and thus interactions among them are expected to affect spin susceptibility at low temperatures and to yield the additional specific heat which is dependent on external magnetic field.

Experiments on Si:P show magnetic field dependent anomaly in specific heat at low temperatures [4] which indicates the freezing of the degrees of spin freedom of localized spins. On the other hand, experimental data on spin susceptibility show Curie-type behavior in the same temperature range and no indication of interactions among localized spins [5,6]. In this paper we first show that this can be understood as the effect of two different kinds of spin-dependent interstate interactions both of which are due to electron-electron interactions among singly occupied states.

In the second part we investigate the effect of the intrastate interaction on the variable range hopping conduction which is also the characteristic phenomenon in the intermediate concentration regime of doped semiconductors. In particular it is shown that the magnetoresistance in the variable range hopping conduction becomes positive in the presence of the intrastate interaction, if the localization lengths for up- and down-spin states are assumed to be the same.

2. Formulation

In the *transfer-diagonal* representation, the Hamiltonian of the system is given by [2,3]

$$H = \sum_{\alpha\sigma} \varepsilon_\alpha \hat{n}_{\alpha\sigma} + \frac{1}{2}\sum_{\alpha\sigma} U_\alpha \hat{n}_{\alpha\sigma}\hat{n}_{\alpha-\sigma} + \frac{1}{2}\sum_{\alpha\beta\gamma\delta}{}' \sum_{\sigma\sigma'} U_{\alpha\beta\gamma\delta} C^+_{\alpha\sigma} C^+_{\beta\sigma'} C_{\delta\sigma'} C_{\gamma\sigma} , \tag{1}$$

where $\hat{n}$'s are number operators, C^+'s and C's are creation and annihilation operators, α, β, γ and δ denote the Anderson-localized states, and σ and σ' denote the electron spins. The first term in (1) is the one-electron Hamiltonian and the second is the intrastate correlation energy. The last term is the remaining electron-electron interactions. $\sum'$ is the summation over α, β, γ and δ excluding the case for $\alpha=\beta=\gamma=\delta$.

The eigenstate $|\alpha>$ is expressed as a linear combination of the Wannier states:

$$|\alpha> = \sum_i v_{i\alpha}|i> \tag{2}$$

with a unitary matrix $v_{i\alpha}$. For the Anderson-localized states, we assume that the wavefunction falls off with an exponential envelope function. Thus the coefficient in (2) can be written as

$$v_{i\alpha} = \tilde{v}_{i\alpha}\exp[-|\vec{r}_i-\vec{r}_\alpha|/\xi_\alpha] \tag{3}$$

where $\tilde{v}_{i\alpha}$ represents a random phase factor and $\vec{r}_\alpha$ the center of the localized state $|\alpha>$. The localization length ξ_α is a function of ε_α, and is considered to behave like

$$\xi_\alpha/a_B \propto |1-\varepsilon_\alpha/E_c|^{-\eta} \tag{4}$$

in the vicinity of the mobility edge, E_c, with a critical index η, where a_B is the effective Bohr radius. Bearing the form (3) in mind, we can estimate the magnitudes of U_α and $U_{\alpha\beta\gamma\delta}$, and it can be shown [2,3] that the intrastate interaction, U_α, is dominant for localized states and that the next largest terms are of the form $U_{\alpha\alpha\alpha\beta}$, $U_{\alpha\beta\beta\alpha}$, and $U_{\alpha\beta\alpha\beta}$. The essential point in estimating the magnitudes of these terms is that, when two localized states α and β are located with the distance between their localization centers equal to R, the matrix elements of U involving α and β acquire factors such as $\exp[-(\xi_\alpha^{-1}+\xi_\beta^{-1})R]$. We can estimate their probable magnitudes by an appropariate probability distribution for the localization centers, and we can show that U_α is dominant unless E_F is in the immediate vicinity of E_c.

The dominant term, U_α, is estimated [7] to be

$$U_\alpha \sim U_0(a_B/\xi_\alpha)^3, \tag{5}$$

where U_0 is the correlation energy within a Wannier state. If we use (4), the energy dependence of $U_\alpha \equiv U(\varepsilon_\alpha)$ is given by

$$U(E) \propto |E_c - E|^{3\eta}, \tag{6}$$

so that U_α vanishes as ε_α approaches the mobility edge E_c.

Thus, unless E_F is very close to E_c, we can retain only the intrastate correlation, U_α, to the first approximation, and we are left with the Hamiltonian

$$H = \sum_{\alpha\sigma} \varepsilon_\alpha \hat{n}_{\alpha\sigma} + \frac{1}{2}\sum_{\alpha\sigma} U_\alpha \hat{n}_{\alpha\sigma}\hat{n}_{\alpha-\sigma}. \tag{7}$$

On basis of this Hamiltonian, the occupation number $n_{\alpha\sigma} \equiv \langle \hat{n}_{\alpha\sigma} \rangle$ is readily given by

$$n_{\alpha\sigma} = f(\varepsilon_\alpha)[1+f(\varepsilon_\alpha)-f(\varepsilon_\alpha+U_\alpha)]^{-1}, \tag{8}$$

where $f(E)$ is the Fermi distribution function. The chemical potential μ is given by the condition

$$N_e = \int NP(E)2f(E)[1+f(E)-f(E+U(E))]^{-1}dE, \tag{9}$$

where N_e is the total number of electrons, N the total number of localized states, and $P(E)$ the probability distribution function for ε_α. For simplicity, we employ a rectangular form for $P(E)$ as

$$\begin{aligned} P(E) &= 1/W \quad \text{for } 0 \le E \le W, \\ &= 0 \quad \text{otherwise.} \end{aligned} \tag{10}$$

In order to take into account the explicit α dependence of U_α, we employ the form of (6) for the entire range of ε_α. Here we assume that $E_c = W$ and $\eta = 0.6$ [7]. Thus we have

$$\begin{aligned} U(E) &= \bar{U}(3\eta+1)(1-E/W)^{3\eta}, \\ \bar{U} &= \frac{1}{W}\int_0^W U(E)dE. \end{aligned} \tag{11}$$

In order to solve (9) with (10) and (11), we need information on N_e and N. In real systems there exist extended states and conduction band states as well as localized states, so that N_e does not necessarily coincide with N. The metal non-metal transition then corresponds to the situation in which μ touches E_c. In this paper we just show the results for the case of $N_e = N$.

We note that the energy level of the second electron on a site which is already occupied by an electron is $\varepsilon_\alpha + U(\varepsilon_\alpha)$, so that the density of second electron levels, $D_2(E)$, is given in our model by

$$D_2(E+U(E)) = |D(E+U(E))/dE|^{-1}NP(E). \tag{12}$$

An example of the profile of the density of states $D_2(E)$ as well as $P(E)$ are shown in Fig.1 [2,3]. In the figure, regions corresponding to singly and

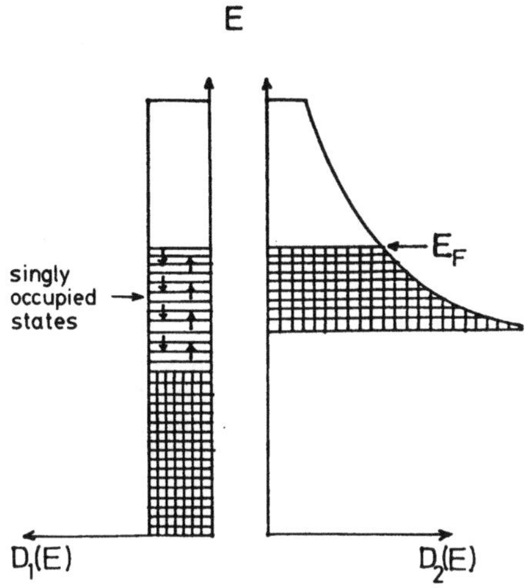

Fig.1 The density of single-particle energy levels $D_1(E) \equiv NP(E)$ and that of second electron levels, $D_2(E)$ for $W/\bar{U}= 6$.

doubly occupied states are shown by hatched and double-hatched regions, respectively. As is seen from (12), $D_2(E)$ takes a large value for an energy for which the derivative $|d(E+U(E))/dE|$ is small. As a consequence, $D_2(E)$ becomes large around $E= E_F$ for $W/U \sim 5$. Another characteristic feature of the density of states in the presence of intrastate correlation is the co-existence of singly and doubly occupied states at and just below the Fermi energy, as mentioned before. This leads to the occurrence of a Curie-type behavior in spin susceptibility at low temperatures where the temperature is much lower than $\bar{U}$ ($\bar{U} > k_BT$). Namely, we obtain [2,3]

$$\chi \sim \mu_B^2 N_s/(k_BT), \tag{13}$$

where N_s is the number of singly occupied states.

3. Spin-Dependent Interstate Interactions

Let us now consider the effect of interstate interactions. If the localization centers are randomly distributed, the largest and the next largest of $U_{\alpha\beta\gamma\delta}$'s in (1) are of the types of $U_{\alpha\alpha\alpha\beta}$ and $U_{\alpha\beta\beta\alpha}$ respectively [2,3]. The latter leads to direct exchange interaction between localized spins with ferromagnetic sign which is due to the same mechanism as Hund's rule in atoms. The former $U_{\alpha\alpha\alpha\beta}$ also leads to interaction between localized spins, because it can mix the configuration of a doubly occupied singlet state into the configuration in which two electrons occupy, for example, α and β states if the spins of α and β states are antiparallel, and thus lowers the energy by this configuration mixing. We call this antiferromagnetic interaction *kinetic-type exchange interaction*. Since both direct and kinetic-type exchange interactions act between spins of singly occupied states, the spin Hamiltonian for states α and β can be written as

$$H_{\alpha\beta} = -2J_{\alpha\beta}(r_{\alpha\beta})\vec{S}_\alpha \cdot \vec{S}_\beta \tag{14}$$

where $\vec{S}_\alpha$ and $\vec{S}_\beta$ are the spins of states α and β with $S_\alpha=S_\beta=1/2$, and $r_{\alpha\beta}$ is the distance between localization centers of states α and β. The exchange interaction $J_{\alpha\beta}$ takes positive or negative sign depending on the relative strength of direct and kinetic-type exchanges. We assume following form:

$$J_{\alpha\beta} = J_0 \exp[-2r_{\alpha\beta}/\xi], \tag{15}$$

where ξ represents the localization length of Anderson-localized states which is here assumed not to depend on states α and β.

4. Spin Pair Model

In order to calculate the specific heat and magnetization of the spin system of singly occupied states, we take the following model. We assume that each localized spin of singly occupied state forms a pair with its nearest neighbor, so that the whole system can be regarded as an ensemble of spin pairs. By the interaction in (14), each pair splits into S=1 triplet state and S=0 singlet state with energy difference $J_{\alpha\beta}$. Thus the free energy of the system is expressed as

$$F = -\sum_{\text{pairs}} k_B T \ln Z_{\text{pair}}(J_{\alpha\beta}), \tag{16}$$

where $Z_{\text{pair}}(J_{\alpha\beta})$ is the partition function of the spin pair system. Denoting the probability distribution function of random values of $J_{\alpha\beta}$ by P(J), (16) can be written as

$$F = -(N_s/2)k_B T \int dJ P(J) \ln Z_{\text{pair}}(J). \tag{17}$$

P(J) is determined, for example, by using (15) and by taking Poisson distribution for localization centers, where equal magnitude for $|J_0|$ is assumed for both positive and negative $J_{\alpha\beta}$. Then the parameters we have are $|J_0|$, the ratio of the number of ferromagnetic pairs ($J_{\alpha\beta} > 0$) to that of the antiferromagnetic pairs ($J_{\alpha\beta} < 0$) N_+/N_-, the number of singly occupied states N_s, and the localization length ξ.

5. Magnetic Field Dependence of Specific Heat Anomaly, and Magnetic Susceptibility

From the free energy in (16), we can now calculate the specific heat and the susceptibility as functions of temperature T and magnetic field H for various values of the above parameters. Calculated results are shown in Fig.2 for the case of ferromagnetic or antiferromagnetic pairs only. The susceptibility χ is small for antiferromagnetic case, while it is approximately of Curie type for ferromagnetic case but with slight deviation from the Curie law due to the existence of singlet excited state. The specific heat is quite small at H=0 for ferromagnetic pairs because the ground state is triply degenerate so that there is residual entropy at T=0. It then rises sharply with external magnetic field and the peak shifts to higher temperature with increasing field. For antiferromagnetic pairs the specific heat curve is insensitive to magnetic fields.

Although the calculated susceptibility for ferromagnetic pairs can explain the Curie-like behavior of susceptibility, the calculated results of specific heat give too sharp rise with increasing magnetic fields and are not consistent with observed behavior. Good fit to experimental results of

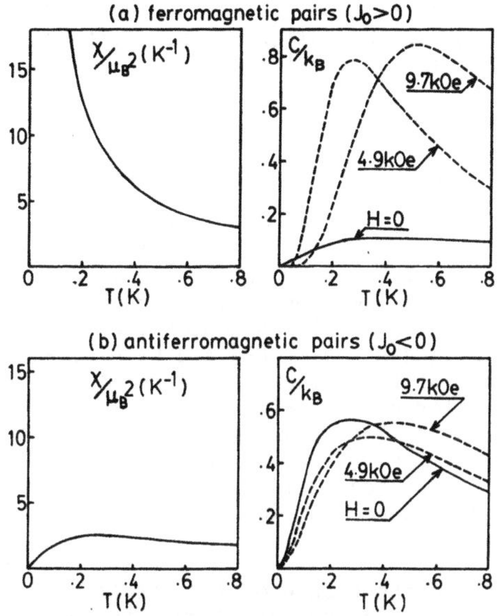

Fig.2 Spin susceptibility χ and specific heat anomaly C per pair of singly occupied states calculated for the ensembles of ferromagnetic pairs (a) and of antiferromagnetic pairs (b).

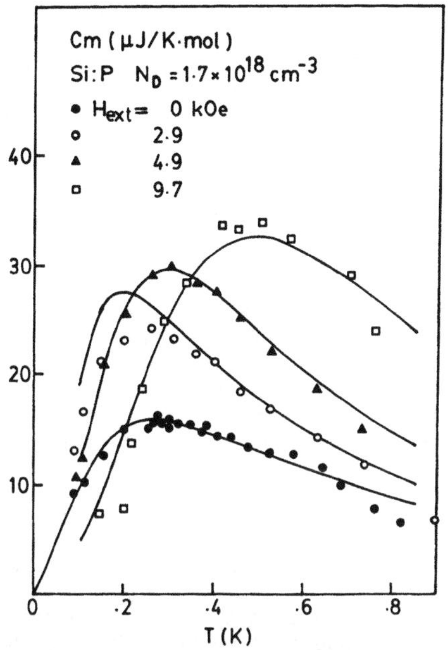

Fig.3 Calculated results of magnetic field dependent specific heat anomaly for donor concentration of $1.7\times10^{18}cm^{-3}$ in the Si:P system (solid line). ●, ○, ▲, and □ are experimental data taken by N. Kobayashi et al. [4] for various magnetic fields.

both excess specific heat and susceptibility was obtained by mixing equal number of ferromagnetic and antiferromagnetic pairs into each other, i.e. $N_+/N_-= 1$. Calculated results of specific heat for this case are shown in Fig.3 together with the experimental data on Si:P ($N= 1.7\times10^{18}cm^{-3}$) by N. Kobayashi et al. [4]. The values of parameters for the best fit to experimental data are $J_0= 5.1K$, $N_+/N_-= 1$, $N_s = 5.6\times10^{17}cm^{-3}$, and $\xi= 81Å$.

Though our model of random spin pairs is simple, the best fit values of parameters are reasonable. J_0 is of the order of the intrastate interaction, and both localization length ξ and the number of singly occupied states ($N_s/N= 33\%$) are also consistent with earlier estimates [2,3] which have been made in explaining the observed T linear specific heat and magnetic susceptibility in the relatively high temperature region.

6. Magnetoresistance in Variable Range Hopping Conduction

In previous sections we have shown that the intra- and inter-state interactions play an important role in the Anderson-localized states. In this section we investigate the effects of these interactions on the variable range hopping conduction, in particular on the magnetoresistance.

As mentioned in §2, there exist three types of electronic states in the presence of the intrastate interaction: unoccupied state(UO state), singly occupied state(SO state) and doubly occupied state(DO state). Corresponding to these three electronic states, there appear the following four different kinds of hopping processes which are shown in Fig.4: (1) hopping processes from SO state to UO state, (2) those from SO to SO state, (3) those from DO to UO state, and (4) those from DO to SO state.

Taking account of these four kinds of hopping processes, we calculate the d.c. conductivity in the presence of a magnetic field for the regime of the variable range hopping conduction. In this calculation we first reduce the problem of magnetoresistance to that of the conduction in a random network of resistors. Then the percolation method developed by Pollak [8] is applied to the calculation of magnetoresistance over a low temperature range. As regards the intrinsic hopping rates between two localized centers, we assume the forms similar to those given by Miller and Abrahams [9] and by Ambegaokar, Halperin and Langer [10]. In the presence of the intrastate correlation, however, a doubly occupied state has a different spatial extension for the envelope functions of two electrons. When the localization length of a state before hopping and that after hopping are different like the case of hopping from DO state to UO state, the larger localization length determines the hopping rate, and this has been taken into account in deriving the intrinsic hopping rates.

By neglecting the spin flip processes and the change of localization lengths due to a magnetic field, the total transition rate from i-th to j-th centers of localization in a magnetic field, Γ_{ij}, can be expressed as

$$\Gamma_{ij} = \sum_{\sigma=\pm 1} 1/(Z_i Z_j)\Big\{\exp[-\beta(E_i + \sigma\mu_B H)]\gamma_{ij}(1) + \exp[-\beta(E_i + E_j)]\gamma_{ij}(2) + \exp[-\beta(2E_i + U)]\gamma_{ij}(3) + \exp[-\beta(2E_i + E_j + U - \sigma\mu_B H)]\gamma_{ij}(4)\Big\}, \tag{18}$$

where $\beta = 1/(k_B T)$, $\gamma_{ij}(\ell)$ represents an intrinsic hopping rate for the above-mentioned four kinds of processes (ℓ=1,2,3 and 4), and Z_i the grand partition function in a magnetic field H which is given by

$$Z_i = 1 + 2\cosh(\beta\mu_B H)\exp(-\beta E_i) + \exp[-\beta(2E_i + U)]. \tag{19}$$

When a magnetic field is applied, the spins of SO states become parallel to the field. Therefore, the probability to find a pair of SO states whose spins are antiparallel becomes smaller with increasing a magnetic field. Thus the hopping processes from SO to SO states which occurs for the spins being antiparallel are suppressed by a magnetic field as far as spin flip processes are neglected. The processes from DO to UO states are connected with those from SO to SO states by the relation of detailed balance. Thus the processes from DO to UO states are also suppressed.

This suppression of hopping processes by a magnetic field gives rise to a positive magnetoresistance. When the suppression of two types of hopping processes, i.e. (2) and (3) in Fig.4, is complete, only the processes from SO to UO states and those from DO to SO states, (1) and (4) in Fig.4, contribute to the hopping conduction. Thus the magnetoresistance saturates for a magnetic field stronger than a certain value.

Explicit calculations for the magnetic field dependence of resistivity have been carried out numerically neglecting the state-energy dependence of the localization lengths and of the intrastate correlation energy, and the

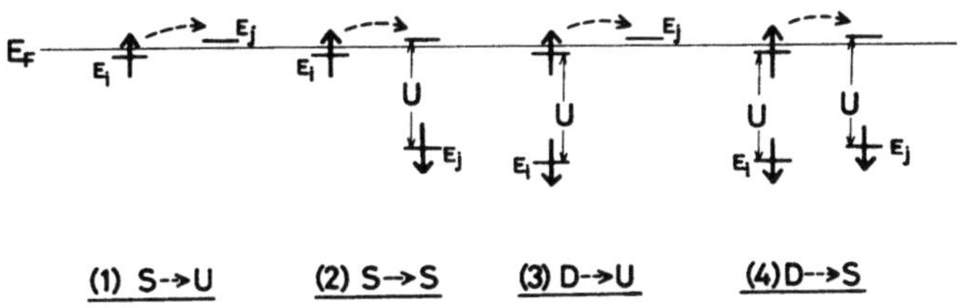

Fig.4 Four kinds of hopping processes in the presence of the intrastate interaction.

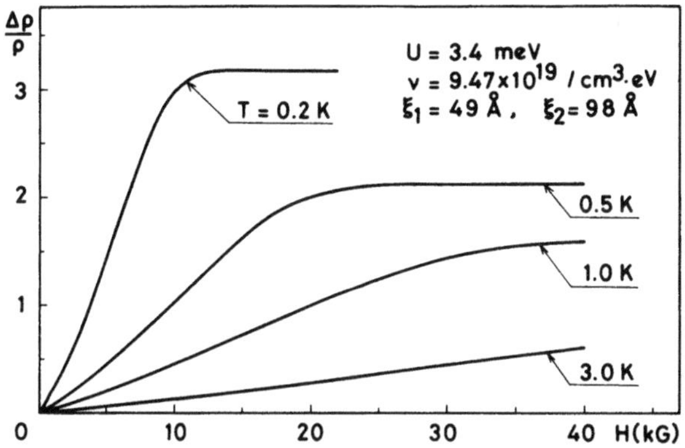

Fig.5 Magnetoresistance as a function of magnetic fields for various temperatures. Only the positive part is taken into account.

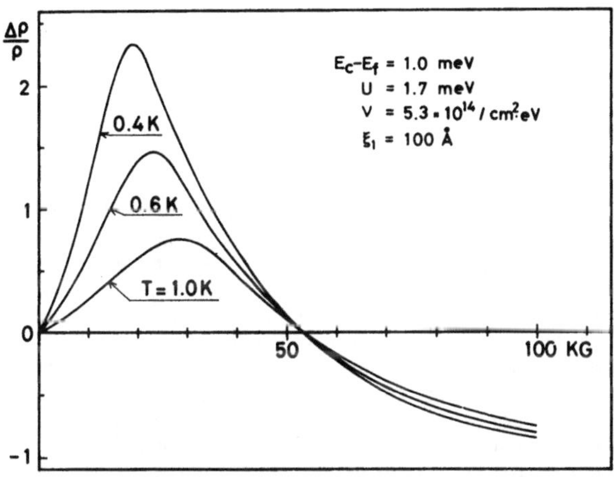

Fig.6 Calculated magnetoresistance of two-dimensional 1T-TaS_2 systems, where both of positive and negative parts in magnetoresistance are taken into account.

result is shown in Fig.5 as a function of a magnetic field for various temperatures. In this calculation the values of the parameters ξ_1, ξ_2, ν and U are chosen from those of Si:P for the impurity concentration of 1.7×10^{18} cm^{-3} [2], where ξ_1 and ξ_2 are the localization lengths of SO and DO states, respectively, ν the density of states at the Fermi level, and U the constant intrastate correlation energy. In this figure we can clearly see that the magnetoresistance increases and then saturates with increasing a magnetic field. The saturation value of magnetoresistance in strong magnetic fields is given by

$$\Delta\rho(H)/\rho(0) = \exp\Big[(13.2/\nu k_B T)^{1/4}$$

$$\times \{(\xi_1^3+\xi_2^3)^{1/4}/(\xi_1^6+\xi_2^6)^{1/4} - (\xi_1^3+3\xi_2^3)^{1/4}/[(\xi_1^3+\xi_2^3)^2+4\xi_2^6]^{1/4}\}\Big] - 1. \quad (20)$$

The saturation value increases when the temperature T or the density of states ν decreases. Furthermore, when the localization lengths decrease keeping a ratio ξ_1/ξ_2 with a constant value, a saturation value also increases.

As an effect of a magnetic field on the transition rate, we so far considered only the change of the occupation probability of each state due to the magnetic field. In addition to this, we have also to take into account the effect of the state-energy dependence of envelope functions on the magnetoresistance. Fukuyama and Yosida [11] investigated this effect and showed that this gives rise to negative magnetoresistance. Combining these two effects, the magnetoresistance can be written conventionally in the following form:

$$\Delta\rho(H)/\rho(0) = N(H) + P(H), \quad (21)$$

where N(H) and P(H) represent negative and positive parts of the magnetoresistance, respectively.

7. Application to 1T-TaS_2 and Its Alloy Systems

Let us calculate the magnetoresistance of two-dimensional disordered systems, with the purpose of explaining peculiar features of magnetoresistance observed in 1T-TaS_2 and its alloy systems. Assuming that the localization lengths depend on the state-energy E_i of i-th state in the following way

$$\xi_1(E_i) = \xi_2(E_i)/2 \propto (E_c - E_i)^{-1}, \quad (22)$$

and choosing the values of the parameters to be (E_C-E_F) = 1.0 meV, U = 1.7 meV, ν = 5.3×10^{14}/cm^{2}eV and $\xi_1(E_F)$ = 100A, we have calculated the magnetoresistance. The result is shown in Fig.6. It is seen from this figure that the magnetoresistance is positive in low magnetic fields and then becomes negative in the region of high magnetic fields. The magnitude of positive part decreases with increasing temperatures. These features are consistent with experimental results [12,13], but in experiments only negative magnetoresistance appears below about 1 K for 1T-TaS_2. This feature does not agree with the theoretical result in Fig.6. This discrepancy, however, can be explained by the spin-dependent interstate interactions between Anderson-localized states. Namely, in the presence of interstate interactions of the form (14), the spins of the SO states which interact by the antiferromagnetic exchange interactions do not become parallel to the magnetic field but still tend to keep the spin singlet state if the magnetic field is weak. Thus the suppression of hopping processes (2) and (3) in Fig.4 by the magnetic field may not occur, and the positive contribution in (21) disappears. In this circumstance, when the temperature is lower than a certain temperature which is of the order of the average interstate interaction, the magnetoresistance becomes negative even in low fields. Recently the specific heat anomaly has been observed in 1T-$TaS_{2-x}Se_x$ [14]. Analyzing this observed result in terms of the present theory mentioned in §4, we can estimate the magnitude of the average interstate interaction to be about 1 K. Thus it can be understood why only the negative part appears in magnetoresistance of 1T-TaS_2 below about 1 K.

Acknowledgments

We would like to thank Dr. Yoshichika Ōnuki, Mrs. Rumiko Inada, Prof. Seiichi Tanuma, Mr. Yutaka Nishio, Dr. Norio Kobayashi, and Prof. Yoshio Muto for showing us their experimental data prior to publication and for helpful discussions. We also acknowledge many stimulating conversations with Prof. Yasutada Uemura and Prof. Shunichi Kobayashi.

References

1 see reviews for example by H. Kamimura in "The Metal Non-metal Transition in Disordered Systems", edited by L.R. Friedman and Tunstall (Edinburgh: SUSSP Publications), p.327; and related references therein.
2 E. Yamaguchi, H. Aoki and H. Kamimura, J. Phys. C 12, 4801 (1979).
3 H. Kamimura, Phil. Mag. B 42, 763 (1980).
4 N. Kobayashi, S. Ikehata, S. Kobayashi and W. Sasaki, Solid State Commun. 32, 1147 (1979).
5 K. Andres, Bull. Am. Phys. Soc. 24, 262 (1979).
6 T. Ema, Master Thesis, University of Tokyo, 1980 (unpublished).
7 H. Kamimura, to be published in Supplement of Progress of Theoretical Physics (1982).
8 M. Pollak, J. Non-Cryst. Solids 11, 1 (1972).
9 A. Miller and E. Abrahams, Phys. Rev. 120, 745 (1960).
10 V. Ambegaokar, B.I. Halperin and J.S. Langer, Phys. Rev. B 4, 2612 (1971).
11 H. Fukuyama and K. Yosida, J. Phys. Soc. Japan 46, 102 (1979).
12 Y. Ōnuki, R. Inada and S. Tanuma, Physica 99B, 177 (1980).
13 N. Kobayashi and Y. Muto, Solid State Commun. 30, 337 (1979).
14 Y. Nishio, N. Kobayashi and Y. Muto, private communications.

Extended Effective Medium Theory for Conductivity

F. Yonezawa, I. Webman* and M.H. Cohen*

Department of Physics, Keio University
3-14-1 Hiyoshi, Kohoku-ku, Yokohama 223, Japan

*Exxon Research and Engineering Company
P.O. Box 45, Linden, NJ 07036, USA

1. Introduction

The problem of the Anderson localization has been studied both analytically and numerically using various methods and theoretical techniques. The central interest for the last few years is concerned with the question how far we can go by applying the concept of scaling and the renormalization group theory to this problem. Now that the attempts along this line are still in progress and that several important features are controversial, it will be worth trying some other methods as well.

In this connection, we can note the idea of the effective medium approximation (hereafter referred to as the EMA) which has been extremely successful in many fields of physics [1]. For instance, the mean-field or molecular-field approximation for phase transitions and the coherent potential approximation (abbreviated as the CPA) are the different versions of the EMA. The idea is also employed in the field of electrical transport in inhomogeneous media and on random networks. Note that the percolation problem is a special example of this field and regarded as a classical analogy of the Anderson problem. Although the percolation theory deals with classical particles while the Anderson localization is concerned with the absence of quantum diffusion due to disorder, it is nonetheless instructive to see what we can learn from the problem of classical transport in random systems.

With this in mind, the purpose of the present article is first to show why the EMA works so well and secondly to investigate how we can improve the EMA.

2. Effective Medium Approximation

Suppose we have an infinite network of resistors as illustrated by Fig.1(a) or Fig.2 where the lattice itself is regular but the conductance of the resistor located on the bonds of the lattice varies from bond to bond. Now, the question is, what is the average conductance of the whole sample?

The idea of the effective medium is first to introduce a hypothetical regular network as shown in Fig.1(b) where the conductance of the resistor on each bond is of the same value throughout the sample. We assume that a current I flows in the network via vertex k and flows out of vertex l. Because of the lattice symmetry and of Kirckhoff's law, it is easy to see that a current I/z flows into each bond connected to site k and that a cur-

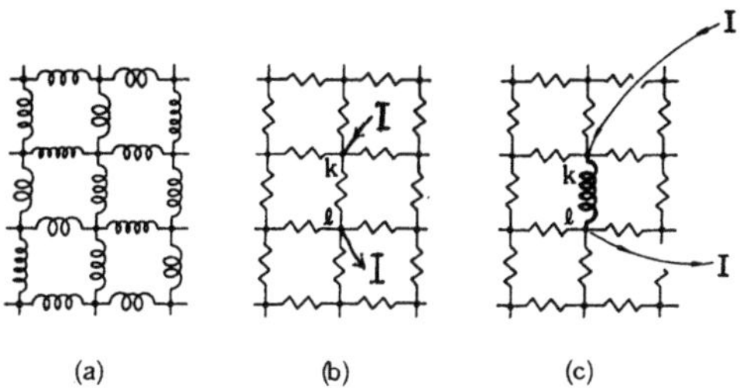

Fig.1 A schematic illustration of the EMA for conductivity on a random resistor network. (a) A network composed of random resistors; (b) a regular network with an equivalent resistor on each bond; and (c) the idea of the EMA

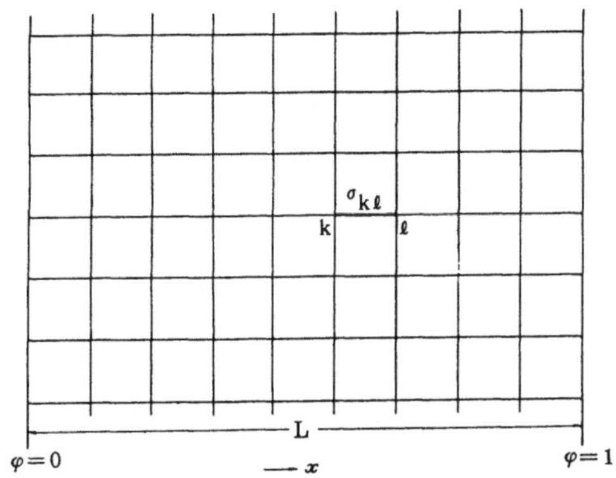

Fig. 2 A random resistor network

rent I/z flows out of each bond connected to site 1, where z is the coordination number. This implies that a current from k to l is $i_M = 2I/z$. The conductance of a resistor on each bond is assumed to be σ_M. Then the conductance of a resistor on bond kl is also σ_M, while we denote by σ_{res} the conductance between k and l through other paths than bond kl. Then we can write

$$\frac{2I}{z\sigma_M} = \frac{I}{\sigma_M + \sigma_{res}} , \tag{1}$$

which gives $\sigma_{res} = \sigma_M(z/2-1)$.

Now, let us study the system as illustrated in Fig.1(c) where the conductance on each bond is the same except for the resistor on bond kl whose conductance is taken to be random. The equation for this case is

$$\frac{i}{\sigma_{kl}} = \frac{I - i}{\sigma_{res}} , \tag{2}$$

where i is the current through the bond kl. Thus, we have $i = I\,\sigma_{kl}/(\sigma_{kl} + \sigma_{res})$. The EMA is to require that the current i on bond kl be equal to i_M on the average. Namely, $\langle i \rangle_{AV} = i_M$ where the brackets $\langle --- \rangle_{AV}$ denote the average over σ_{kl}. This requirement leads to

$$\left\langle \frac{\sigma_{kl} - \sigma_M}{\sigma_{kl} + \sigma_M(z/2-1)} \right\rangle_{AV} = 0 . \tag{3}$$

It is interesting to note that, for the case of a simple-cubic lattice with z = 6, (3) is identical with the result of the EMA for a continuous and inhomogeneous media of three dimensions [1], [2].

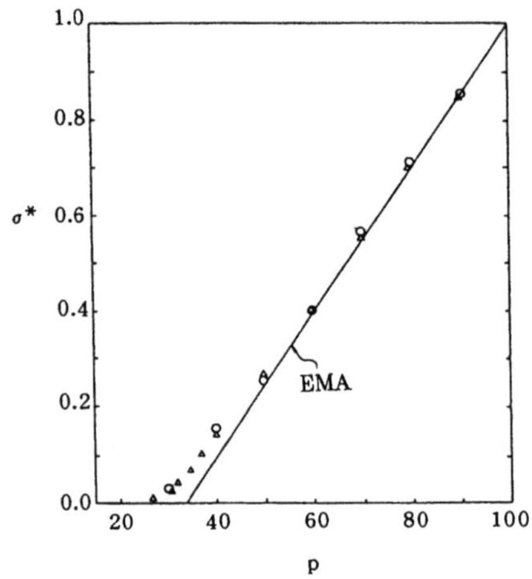

Fig.3 The average conductivity of a random network on a simple-cubic lattice composed of two kinds of resistors characterized by conductivities $\sigma_1 = 1$ and $\sigma_2 = 0$ with respective concentration p and 1-p. The solid line represents the result of the EMA while data points indicate the results of the computer simulation due to KIRKPATRICK [4] for simple cubic networks which range in size from 15x15x15 to 25x25x25. Throughout this paper, the length of each bond is taken to be the unit length, and the words conductance and conductivity are both used according to the context concerned

When the distribution of σ_{kl} is such that $\sigma_{kl} = 1$ or 0 with respective probability p and 1-p, the model reduces to the percolation problem [3]. The EMA equation (3) for a simple-cubic lattice then is expressed as $\sigma_M = (3p-1)/2$. In Fig.3, this result of the EMA is compared to the data obtained from a computer simulation [4]. In spite of the fact that the concept of the EMA is rather simple-minded as explained in the above, the agreement is markedly good except for the concentration region just above the exact threshold $p_c(\text{exact}) \simeq 0.249$. Note that $p_c(\text{EMA}) = 1/3 = 0.3333....$

3. Formulation of the Problem

For the purpose of giving a systematic way to improve approximations step by step beyond the EMA, we write in this section a formulation in terms of the Green's function.

The lattice is assumed to be infinite in the y and z directions. In the x direction, it is bounded by two planes at x = 0 and x = L (see Fig. 2), where we let L go infinity at a later stage.

For each realization of the random lattice, the potentials at the sites $\{\phi_i\}$ obey a set of linear equations which can be recast in the following matrix notation:

$$A \Phi = S , \tag{4}$$

where

$$A = \sum_{kl} |k\rangle \left(\delta_{kl} \sum_{n} \sigma_{kn} - \sigma_{kl} \right) \langle l| \quad , \tag{5}$$

and

$$\Phi = \sum_{i} \phi_i \, |i\rangle \quad . \tag{6}$$

The matrix S represents the potentials at the boundaries. Note that σ_{kl} is nonzero only when k and l are nearest neighbours to each other.

The matrix A can be represented as a sum of a homogeneous part A_M and a part δA which contains the randomness. If we assume that the homogeneous part A_M corresponds to the effective medium as illustrated by Fig.1(b), then we have

$$A_M = \sigma_M \sum_{kl} |k\rangle \left(z\,\delta_{kl} - \Delta_{kl} \right) \langle l| \, , \tag{7}$$

where

$$\Delta_{kl} = \begin{cases} 1 \text{ when k and l are nearest neighbours to each other,} \\ 0 \text{ otherwise.} \end{cases}$$

Then it is easy to see that

$$\delta A = A - A_M = \sum_{kl} |kl\rangle \left(\sigma_{kl} - \sigma_M \right) \langle kl| \quad , \tag{8}$$

where

$$|kl\rangle \equiv |k\rangle - |l\rangle \, , \tag{9}$$

and the summation is over pairs of nearest neighbour sites.

The local field E_{kl} along the (kl) bond is given by

$$E_{kl} = \phi_k - \phi_l = \sum_{i \in B} \langle kl| \, A^{-1} \, |i\rangle \equiv \sum_{i \in B} \langle kl| \, G \, |i\rangle \quad , \tag{10}$$

where the summation is over all sites on the x=L plane. For a homogeneous effective medium, on the other hand, we have

$$E^M_{kl} = \sum_{i \in B} \langle kl| \, A_M^{-1} \, |i\rangle \equiv \sum_{i \in B} \langle kl| \, G_M \, |i\rangle = L^{-1} \, . \tag{11}$$

The effective or average conductivity σ^* is defined as

$$\sigma^* = \frac{\left\langle \sum_{kl} \sigma_{kl} \, E_{kl} \right\rangle_{AV}}{\left\langle \sum_{kl} E_{kl} \right\rangle_{AV}} \quad . \tag{12}$$

The total Green's function G can be expanded in terms of the Green's function for the effective medium $G_M = 1/A_M$, i.e.

$$G = A^{-1} = G_M + G_M \,\delta A\, G_M + G_M \,\delta A\, G_M \,\delta A\, G_M + \ldots . \tag{13}$$

From (12) and (13), we obtain

$$\sigma^* - \sigma_M = \Big\langle \Big[\sum_{\underline{kl}} (\sigma_{\underline{kl}} - \sigma_M) + \sum_{\underline{kl}} \sum_{\underline{mn}} (\sigma_{\underline{kl}} - \sigma_M) \Gamma(\underline{kl},\underline{mn}) (\sigma_{\underline{mn}} - \sigma_M)$$

$$+ \sum_{\underline{kl}} \sum_{pq} \sum_{\underline{mn}} (\sigma_{\underline{kl}} - \sigma_M) \Gamma(\underline{kl},pq) (\sigma_{pq} - \sigma_M) \Gamma(pq,\underline{mn}) (\sigma_{\underline{mn}} - \sigma_M)$$

$$+ \cdots\cdots \Big] \Big\rangle_{AV} \tag{14}$$

where the underline indicates that the bond is parallel to the applied field. Note that

$$\Gamma(kl,mn) = -[G_{km} - G_{kn} - G_{lm} + G_{ln}] \ , \tag{15}$$

$$G_{km} = \langle k | A_M^{-1} | m \rangle \ , \tag{16}$$

where $G_{km} \sigma_M$ is the ordinary lattice Green's function.

It is often more convenient in formulations if we rearrange the perturbation expansion series (14) in terms of t-matrices. Then we have

$$\sigma^* - \sigma_M = \Big\langle \sum_{\mu} t_\mu + \sum_{\mu} \sum_{\neq \nu} t_\mu \Gamma_{\mu\nu} t_\nu + \sum_{\mu} \sum_{\neq \nu} \sum_{\neq \tau} t_\mu \Gamma_{\mu\nu} t_\nu \Gamma_{\nu\tau} t_\tau + \cdots \Big\rangle_{AV} \tag{17}$$

where the t-matrix is defined by (note that μ, ν etc. represent bonds)

$$t_\mu = (\sigma_\mu - \sigma_M)/[1 - (\sigma_\mu - \sigma_M)\Gamma_{\mu\mu}] \ . \tag{18}$$

4. EMA in the Green's Function Formulation

Now that we have tools, we can examine how we improve approximations systematically. In what follows, we take as an example a simple-cubic lattice for which $\Gamma_{\mu\mu} = -1/3\sigma_M$.

[Approximation I]

In analogy with the idea of the CPA, let us require that the t-matrix be zero on the average, i.e.

$$\langle t_\mu \rangle_{AV} = 0 \ , \tag{19}$$

which leads to (3) when z = 6 —— the value for a simple-cubic lattice.

The reason why the EMA works well is the same as that for the CPA. Suppose we represent equations by means of diagrams using the rules for counting equations (see Fig.4). Then, it is easy to see that, among higher-order diagrams associated with each term in the perturbation expansion series (17), all diagrams with one or more independent t-matrices contribute nothing.

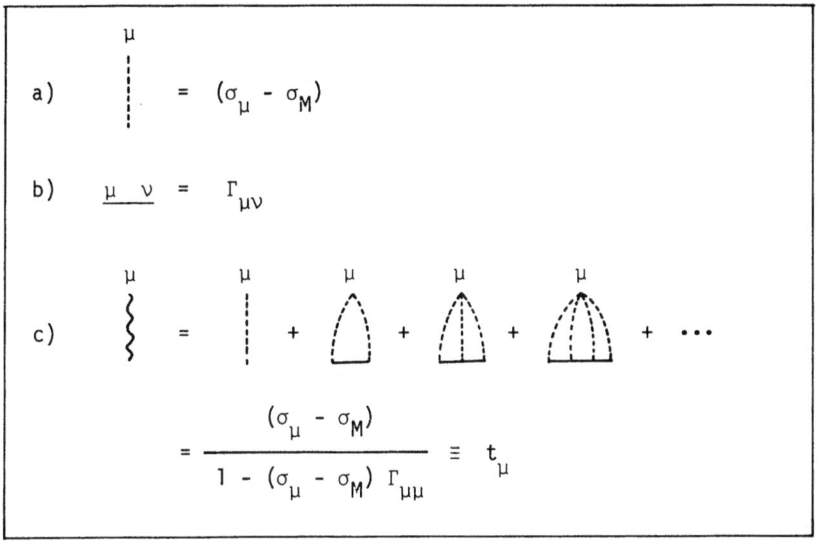

Fig.4 Rules for expressing equations by means of diagrams

5. Double-Bond Cluster EMA

In this section, we consider how we can go beyond the EMA by taking into account some important terms excluded from the EMA.

[Approximation II]

When the t-matrix is required to vanish on the average, the lowest-order term which does not disappear is, as is the case of the CPA, the fourth-order in the t-matrix of the form

$$\sum_{\mu\nu} C^{(4)}_{\mu\nu} = \sum\sum_{\mu\neq\nu} t_\mu \Gamma_{\mu\nu} t_\nu \Gamma_{\nu\mu} t_\mu \Gamma_{\nu\mu} t_\mu \tag{20}$$

which is expressed by a crossed diagram as shown in the second colomn of Fig.5. Since only two bonds are concerned, this kind of diagram is called a double-bond cluster diagram. Apparently, an approximation which is one step better than the EMA is attained by requiring

$$\left\langle \sum_\mu t_\mu + \sum\sum_{\mu\neq\nu} C^{(4)}_{\mu\nu} \right\rangle_{AV} = 0 \,. \tag{21}$$

This gives approximation II in Fig.5 and the improved value for p_c(see Fig. 5).

[Approximation III]

The idea is clearly shown in Fig.5 where approximation III is to require the single-bond diagram (the t-matrix), the fourth-order double-bond diagram and the fifth-order double-bond diagram as a whole be zero on the average. As for the results, see Fig.5

No.	APPROXIMATION	p_0
I	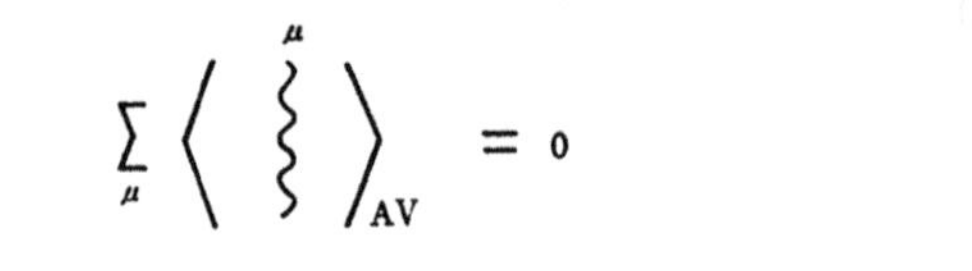	0.3333
II	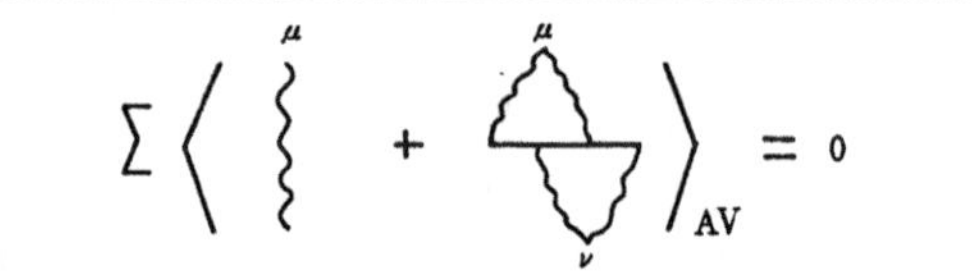	0.316
III	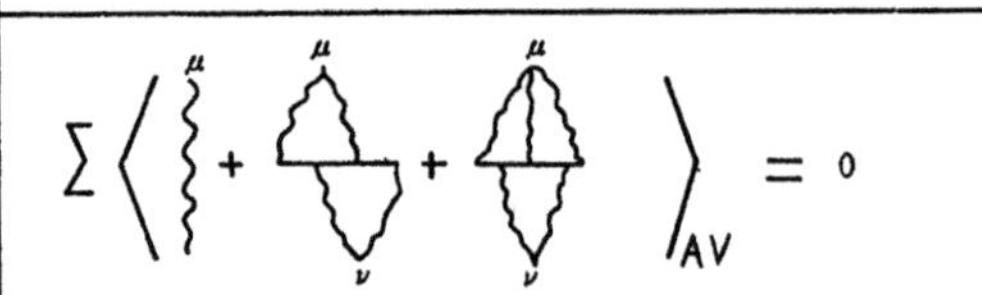	0.304
IV	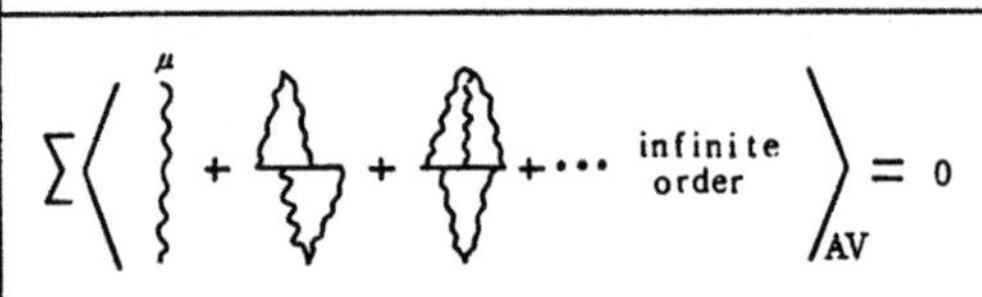	0.296
V	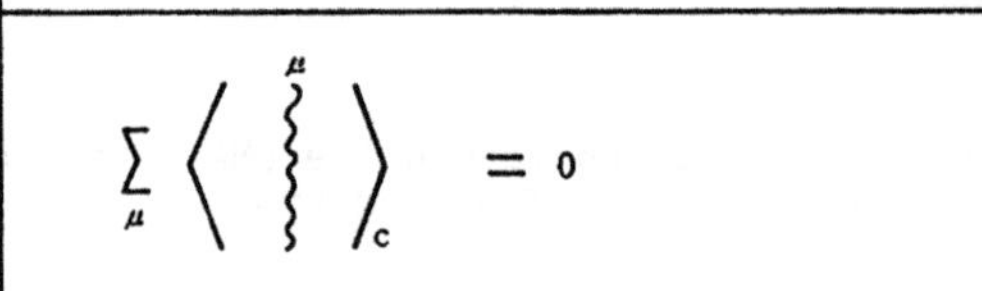	0.283
VI	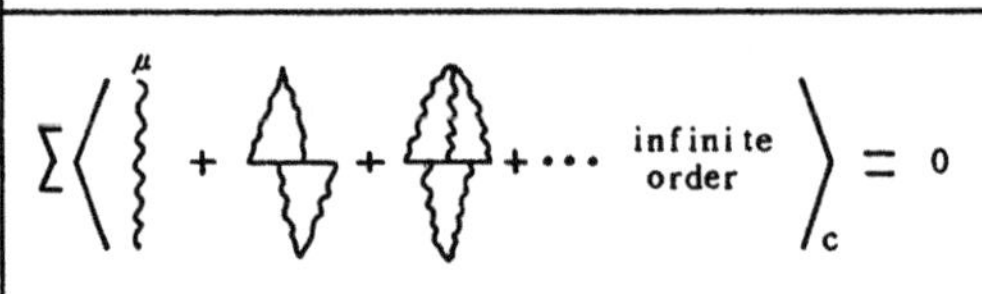	0.258

Fig. 5 Various approximations and their theoretical values for the critical percolation concentration

[Approximation IV]

In this approximation, the single-bond diagram and all the double-bond diagrams are assumed to vanish on the average. Naturally, the terms taken into account are up to the infinite order. Even in this case, the change in the theoretical value for p_c is appreciable but the convergence is slow.

6. Cumulant

The difficulty becomes increasingly serious when we try to include higher-order diagrams than two-bond clusters. In addition to the complexity inherent in higher-order diagrams, the most troublesome feature comes from the fact that the summations over bonds must be taken under the restriction that any pair of succeeding two bonds be different from each other as shown in (17). One excellent way for removing this restriction on the summations is to use the cumulant average $\langle --- \rangle_C$ instead of the ordinary average $\langle --- \rangle_{AV}$. In the cumulant description, the perturbation expansion (17) is rearranged in a form where the summations over bonds can be carried out perfectly independent of one another. Advantageous features attained by using the cumulant average were fully discussed in [5] in connection with the problem of the CPA. Among various merits of the cumulants, the outstanding ones are that the inclusion of higher-order terms becomes comparatively easier and that the mathematical structure of approximations can be analyzed systematically.

Besides these advantages, our problem has an additional merit that $\Sigma_\nu \Gamma_{\mu\nu}$ is identically zero for a simple-cubic lattice when the summation is not restricted. This means that a lot of higher-order diagrams with one or more independent Γ-lines vanish automatically and therefore the qualities of lower-order approximations are much higher than those of the corresponding approximations in the ordinary average description. The examples are given in the following.

[Approximation V]
The single-bond cumulant approximation is obtained by requiring that the t-matrix be zero on the cumulant average (cf. the fifth colomn in Fig.5). The effective conductivity in this approximation is shown in Fig.6(a). The agreement with the results of the computer simulation is remarkably good. Note that the corresponding results have been obtained also for the conductivity in an inhomogeneous media by HORI and YONEZAWA [2].

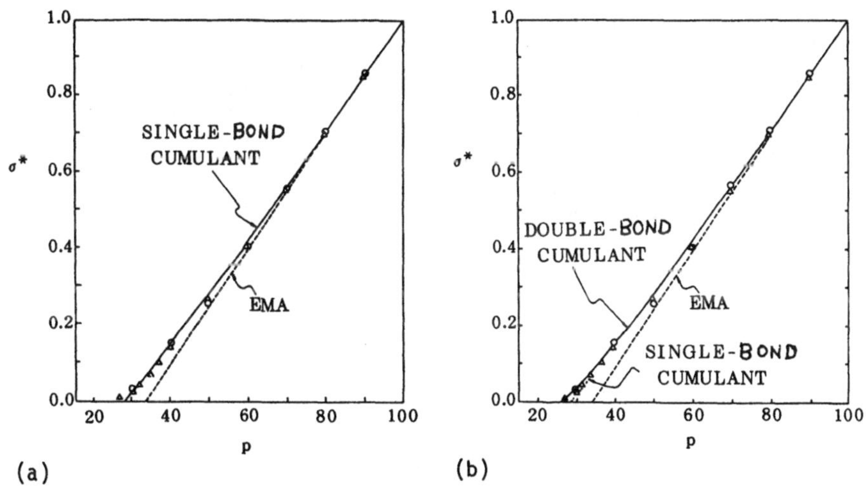

Fig.6 (a) The effective conductivity (the solid curve) within the single-bond cumulant approximation compared to the results of the EMA(the broken line) and of the computer simulation. (b) The effective conductivity within the double-bond cumulant (the solid curve) compared to the results of the EMA (the broken curve), of the single-bond cumulant (the dotted curve) and of the computer simulation

The critical percolation concentration in this approximation is given to be 0.283 which is already better than the best double-bond approximation (IV) which gives 0.296.

[Approximation VI]
The idea of the double-bond approximation in the cumulant is clearly expressed by the diagrams in the last colomn in Fig.5. The obtained results are surprisingly good as shown in Fig.6(b) and by the value of the critical threshold which is 0.258.

7. Discussion

We have given some brief description concerning why the EMA works so well and how we can improve approximations systematically. Because of the limited space, we cannot present a detailed formalism nor a comprehensive analysis. A full paper will be found elsewhere[6].

References

[1] Early references of the effective medium theory are found in; R.Landauer, in Electrical Transport and Optical Properties of Inhomogeneous Medis, eds. J.C.Garland and D.B.Tanner (Americal Institute of Physics, 1978), p.2.
[2] M.Hori and F.Yonezawa, J.Math.Phys. 15, 2177(1974); ibid.16, 352, 365 (1975); J.Phys.C 10, 229 (1977)
[3] Concerning the percolation theory, see for instance the following papers.
S.R.Broadbent and J.M.Hammersley, Proc.Camb.Phil.Soc. 53, 629 (1957).
V.K.S.Shante and S.Kirkpatrick, Adv.Phys. 20, 325 (1971).
D.J.Thouless, Phys.Rep. 12, 93 (1974); in Ill-Condensed Matter, eds. R.Balian, R.Maynard and G.Toulouse (North Holland, Amsterdam, 1979), p.1.
[4] S.Kirkpatrick, Phys.Rev.Lett. 27, 1722 (1971); Rev.Mod.Phys. 45, 574 (1973); in The Properties of Liquid Metals, ed. S.Takeuchi (Taylor and Francis, London, 1973), p.351.
[5] F.Yonezawa, Progr. Theor.Phys. 40, 734 (1968).
[6] F.Yonezawa, I.Webman and M.H.Cohen, submitted to Phys.Rev. (1982).

Part II

Quantized Hall Effect

motion, and $E_N=(N+1/2)\hbar\omega_c$ with $\omega_c=eH/mc$. As is shown by KUBO et al. [17], the coordinate of an electron can be expressed as a sum of relative and center coordinates by $x=\xi+X$ and $y=\eta+Y$ with $\xi=(\ell^2/\hbar)\pi_y$ and $\eta=-(\ell^2/\hbar)\pi_x$.

The Hall conductivity is expressed [17] as

$$\sigma_{xy} = -\frac{nec}{H} + \Delta\sigma_{xy} , \tag{2.3}$$

where n is the electron concentration and $\Delta\sigma_{xy}$ is given by the correlation function of the velocities of the center coordinates X and Y as

$$\Delta\sigma_{xy} = \frac{e^2\hbar}{iL^2} \sum_\alpha < f(E_\alpha) \sum_\beta \mathrm{Re}\,(E_\alpha-E_\beta+i0)^{-2} \times[(\alpha|\dot{X}|\beta)(\beta|\dot{Y}|\alpha) - (\alpha|\dot{Y}|\beta)(\beta|\dot{X}|\alpha)] > , \tag{2.4}$$

with f(E) being the Fermi distribution function, $|\alpha)$ the eigenstate of the Hamiltonian (2.1) with eigenenergy E_α, and L the linear dimension of the system. If $|\alpha)$ is a localized state, we have, for any $|\beta)$,

$$(\alpha|\dot{X}|\beta) = (i\hbar)^{-1}(\alpha|X|\beta)(E_\alpha-E_\beta). \tag{2.5}$$

Therefore, the contribution from the state $|\alpha)$ to $\Delta\sigma_{xy}$ in (2.4) reduces to $\Delta\sigma_{xy}{}^\alpha=f(E_\alpha)ec/H$, where use has been made of $[X,Y]=i\ell^2$. From this result we can derive important general properties at T=0. (i) As long as the Fermi level lies in the localized regime of the energy spectrum, σ_{xy} remains constant independent of n. (ii) If all the states below the Fermi level E_F are localized, we have $\sigma_{xy}=0$ because $\Delta\sigma_{xy}$ exactly cancels -nec/H in (2.3).

Let us now study the limit of strong magnetic field, in which the energy spectrum comprizes separate Landau levels. In this case there is a situaion in which Landau levels are just filled up to the Nth level when E_F lies in the gap between the Nth and (N+1)th levels with $n=N/2\pi\ell^2$. Now it can be shown that (iii) $\Delta\sigma_{xy}=0$, i.e. $\sigma_{xy}=-Ne^2/h$ in such a case. This is readily seen from (2.4) if we note that $f(E_\alpha)$ can be replaced by $f(E_\alpha)[1-f(E_\beta)]$ in that equation. It follows that, in the limit of strong magnetic fields, (iv) σ_{xy} assumes the constant value $-Ne^2/h$ when E_F lies in the localized regime between the Nth and (N+1)th levels according to the properties (i) and (iii). Combining the properties (ii) and (iii), we can also draw an important conclusion that (v) all the states cannot be localized. Using a symmetry property of the Hamiltonian we can further show that

$$\Delta\sigma_{xy}(V,\nu) = - \Delta\sigma_{xy}(-V,1-\nu), \tag{2.6}$$

where ν is the occupation ratio of the Landau level defined by $\nu=2\pi\ell^2n_N$ with n_N being the electron concentration of the level. By this formula we can relate σ_{xy} in the lower edge of a Landau level to that in the upper edge of the level.

The properties (iii)-(v) described above have been derived in the limit of strong magnetic field where matrix elements between different Landau levels in (2.2) can be completely neglected. LAUGHLIN [7] has given an argument which shows that there can be no corrections due to mixings of different Landau levels to the value $-Ne^2/h$. Here, we shall present a slightly different explanation [18]. We consider an L×L system and impose periodic boundary conditions in the y direction. Electrons are assumed to be free to pass through the left (x=0) and right (x=L) edges of the system. The center coordinate X in (2.2) in the absence of an external electric field is $X=-\ell^2 k_y=(2\pi\ell^2/L)\times j$ where j=1, 2, ..., $L^2/2\pi\ell^2$. An electric field E_y is introduced by a time-dependent vector potential A_y which is independent of x and y. Now we increase A_y very slowly from zero. Each time A_y reaches the integer multiple of $(2\pi/L)(\hbar c/e)$, the Hamiltonian of the system becomes identical with that at t=0 (A_y=0) if we perform a gauge transformation by multiplying a phase function which is periodic in the y direction. Let us now assume that the Fermi level lies in a nonzero gap in the spectrum of extended states. After the time Δt, corresponding to the change in A_y by $(2\pi/L)(\hbar c/e)$, the wave function of the system becomes identical with the previous one apart from a simple phase difference. Here, we have used the fact that occupations of localized states are not influenced by the vector potential A_y because localized states do not contribute to current. The only essential change after the time Δt will be that a certain number (p) of electrons disappear through the left edge and the same number of electrons enter the system through the right edge. Therefore, the current will be

$$j_x = -\frac{p(-e)}{L}\frac{Le}{hc}\frac{\Delta A_y}{\Delta t} = -p\frac{e^2}{h}E_y\ , \tag{2.7}$$

which leads to the quantized Hall effect.

3. Method of Numerical Experiments

We consider a system with a finite size L×L. Scatterers are distributed randomly and their potential is given by

$$V(r) = \sum_i v(r-r_i), \tag{3.1}$$

where r_i denotes the position of the ith scatterer. We shall use periodic boundary conditions for the both x and y directions. To this end we construct an infinite periodic lattice in which the L×L system constitutes a unit cell and as the basis in the absence of scatterers we choose

$$\psi_{NX}(r) = M^{-1/2}\sum_m \phi_{NX+mL}(r), \tag{3.2}$$

where $\phi_{NX}(r)$ is the usual Landau wave function for a gauge A=(0,Hx) and M is the total number of cells. This $\psi_{NX}(r)$ is periodic in the y direction for $X=X_j\equiv j\Delta X$ (j=1, 2, ..., N_X) with $N_X=L^2/2\pi\ell^2$ and $\Delta X=2\pi\ell^2/L$. It is periodic also in the x direction except for a phase factor, i.e.

$$\psi_{NX}(x+L,y) = \psi_{NX}(x,y)\exp(-iLy/\ell^2), \tag{3.3}$$

as is desired. We see that $\psi_{NX}(r)$ for $X=X_j$ (j=1, 2, ..., N_X) constitutes an orthonormal basis set.

The matrix element of the potential becomes

$$(NX|V|N'X') = \int dr\ \phi^*_{NX}(r)[V(r)+V(x+L,y)+V(x-L,y)]\phi_{N'X'}(r) + \sum_{\pm}\int dr\ \phi^*_{NX}(r)[V(r)+V(x\mp L,y)]\phi_{N'X'\pm L}(r), \tag{3.4}$$

where couplings among different cells other than between the nearest neighbor cells have been neglected since the system size L is much larger than the cyclotron radius ℓ. The magnetic field is assumed to be strong enough for each Landau level to be treated separately. In this case the energy levels can be obtained by numerically diagonalizing the matrix $(NX|V|NX')$ for each N. We assume equal amounts of attractive and repulsive scatterers with a potential of a Gaussian form:

$$v(r) = \pm (V/\pi d^2) \exp(-r^2/d^2), \tag{3.5}$$

which becomes $\pm V\delta(r)$ in the limit $d \to 0$. For such scatterers we can use a symmetry around the center of each Landau level.

Whether states are localized or extended is determined by a study of the Thouless number g(L) which is defined as the ratio of the shifts $\langle \Delta E \rangle$ of the individual energy levels due to a change in boundary conditions to the level separation $[L^2 D(E)]^{-1}$, where D(E) is the density of states per a unit area. If states are localized, g(L) is expected to decrease with the increase of the system size L reflecting how the wave function decays with the distance of the localization center. For extended states the energy shifts should be of the order of $\hbar/\tau$, where τ is the time for the electron to diffuse away to the boundary, and g(L) is expected to be independent of L.

LICCIARDELLO and THOULESS [16] have shown that the conductivity in two dimensions can be directly related to the Thouless number g(L). The same theory can be applied fairly straightforwardly to the case in strong magnetic fields if we note that the velocity operators v_x and v_y can now be replaced by $\dot{X}$ and $\dot{Y}$ in strong magnetic fields. Under the assumption (i) that the square modulus of the velocity matrix element is uncorrelated with the corresponding energy difference and (ii) that couplings between different sites can be well approximated by a tight-binding model, we have

$$\sigma_{xx} = \frac{1}{4}\frac{e^2}{\hbar}\ g(L)\ . \tag{3.6}$$

The assumption (ii) should be a reasonable one near the spectral edges because couplings are small compared with spacings between levels. It is not valid near the center of Landau levels especially with a higher index, as will be demonstrated below.

The energy shifts ΔE are calculated by replacing periodic boundary conditions by antiperiodic conditions in the y direction. This can be achieved by using $X = X'_j \equiv (j-1/2)\Delta X$ instead of $X = X_j$. Numerical calculations are performed for several different values of N_X and for the lowest three Landau levels (N=0, 1, and 2) in case of short-range delta-potential scatterers and the two levels (N=0 and 1) in case of long-range ($d=2\ell$) scatterers. Numbers of different samples are chosen to be more than 50, 40, 30, 20, and 10 for N_X=36, 64, 100, 144, and 196, respectively. Actual numbers are twice as large as these because of the symmetry property around the center of the Landau level. The linear dimension L is about 15, 20, 25, 30, and 35 in units of ℓ for the above N_X's.

It has previously been shown that for delta-potential scatterers discrete levels which are not affected by scatterers and have the energy E_N appear when their concentration is small [19]. The condition of the appearance is given by $c_i<1$, where c_i is the average number of scatterers contained in the lowest cyclotron orbit, i.e. $c_i=2\pi\ell^2 n_i$ with n_i being the concentration of scatterers. In this paper, therefore, we assume a rather high concentration of scatterers ($c_i=5$). Such singular behavior does not arise in case of scatterers with a finite potential-range and $c_i=1$ is assumed for long-range ($d=2\ell$) scatterers. In the both cases the average number of scatterers which have influence on an electron at the same time is nearly equal.

The resulting energy levels and wave functions for periodic boundary conditions are used also for the calculation of the Hall conductivity of a finite system. It should be noted that the Hall conductivity is not a dissipative quantity and not determined by energy levels near the Fermi level alone. Consequently the expression for the Hall conductivity (2.4) does not contain a singular delta-function which causes serious problems in evaluating the diagonal conductivity σ_{xx} in a finite system. This means, however, that the knowledge of all the energy levels and their wave functions is required for the evaluation of (2.4).

4. Results of Numerical Experiments

In the following the energy will be normalized by the level broadening Γ_N obtained in a self-consistent Born approximation [20]. First we shall present the results for short-range scatterers ($d=0$). In this case Γ_N becomes independent of N, i.e. $\Gamma_N=\Gamma$. Figure 1 shows examples of calculated histograms of the density of states for N=0, 1, and 2. The density of states becomes slightly narrower than the dashed line obtained in a single-site approximation [19]. It can be clearly seen that the tails of the density of states become smaller and the density of states approaches that calculated in the single-site approximation with increasing the level index N. This is in agreement with the result obtained previously by summing up an infinite perturbation in an approximate way [21].

Histograms of the Thouless number $g(L)$ for N=0 are given in Fig.2 and for N=1 and 2 in Fig.3, where the logarithm of $g(L)$ for different L's is plotted against energy. Only the lower half ($E<0$) is shown because of the symmetry. The Thouless number has been calculated only for $N_X=36$, 64, 100, and 144 in case of the lowest Landau level. For N=0, $g(L)$ decreases with increasing L below $E\sim-0.2\Gamma$, while no such tendency is observed above $\sim-0.2\Gamma$. This shows that states are not localized in the energy range between $\sim-0.2\Gamma$ and $\sim0.2\Gamma$ for the present system size. On the other hand, the exponential decay of $g(L)$ with increasing L below $\sim-0.2\Gamma$ or above $\sim0.2\Gamma$ indicates that states are exponentially localized in these energy ranges. In case of N=1 and 2 the clear exponential decrease of $g(L)$ can be observed below $E\sim-0.6\Gamma$ and -0.8Γ, respectively. In the other energy range no such systematic behavior can be seen in the L dependence of $g(L)$.

In the localized regime, we can determine the extent of localized wave functions, $\alpha(E)^{-1}$, by putting $g(L)=g(0)\exp(-\alpha L)$. The resulting α's are plotted against energy in Fig.4 for the lowest three Landau levels. We see that α decreases almost linearly with increasing energy. The existing large fluctuations of the results make impossible the exact determination of the critical exponent s defined by $\alpha(E)\propto|E-E_c|^s$. The straight lines shown in the figure are the least-square fit to the data points. By such a simple extrapolations the "mobility edges" are assigned as $\sim\pm0.2\Gamma$, $\sim\pm0.6\Gamma$, and $\sim\pm0.7\Gamma$ for N=0, 1, and 2, respectively. It should be noticed,

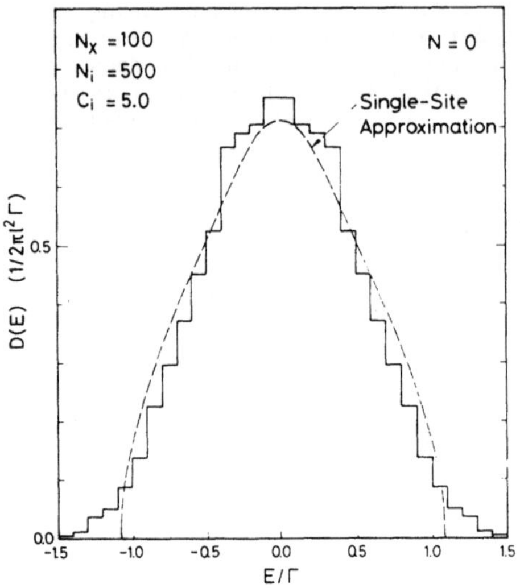

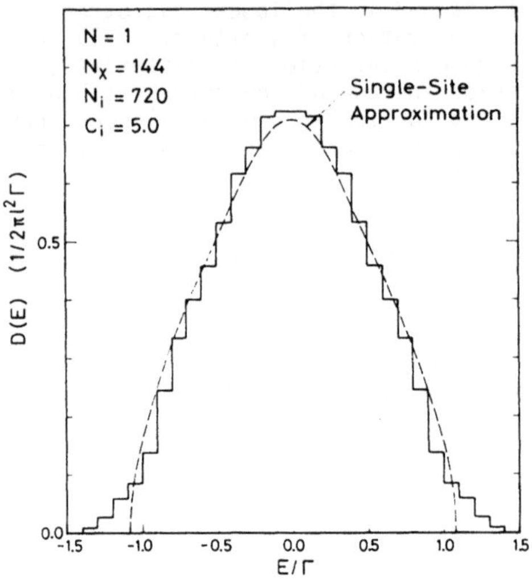

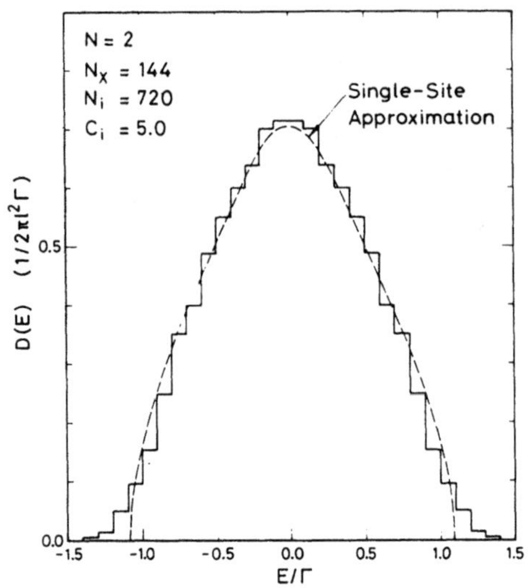

Fig.1 Calculated histograms of the density of states of the N=0, 1, and 2 Landau levels for short-range delta-potential scatterers. The result in a single-site approximation is given by the dashed line.

however, that the assignment is not a final one and the resulting "mobility edges" should be regarded as a threshold energy below which the clear exponential localization is observed for the system size considered.

Figure 5 shows the Hall conductivity of the lowest two Landau levels N=0 and 1. For N=0, the absolute value of the Hall conductivity remains very close to zero below $\sim -0.2\Gamma$, increases almost linearly with energy, and then takes the value e^2/h above $\sim 0.2\Gamma$. For the N=1 Landau level, only the contribution from the N=1 level is shown, i.e. the actual Hall conductivity is given by adding $-e^2/h$, which is the contribution of the N=0 level. The results for the two different system sizes (N_X=64 and 144) are given for

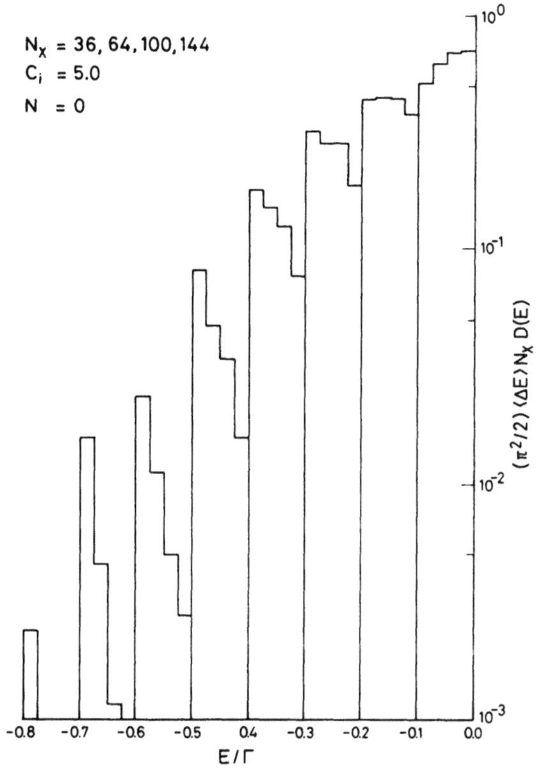

Fig.2 Calculated Thouless number g(L) of the lowest Landau level (N=0) for short-range delta-potential scatterers. Within each energy bin, the behavior with increasing system size L or N_X is plotted from the left.

N=1. In principle, we could discuss whether states are localized or not by a study of the dependence of the Hall conductivity on the system size, which, however, requires much longer computational time. The calculated Hall conductivity depends on N_X most strongly around $E\sim\pm0.4\Gamma$ but not below $\sim-0.7\Gamma$ or above $\sim0.7\Gamma$. It is interesting to note that the Hall conductivity starts to deviate from the integer multiple of $-e^2/h$ near the "mobility edge" obtained in Fig.4. If we consider the fact that the system size is not large enough, this suggests that the actual mobility edge should lie closer to the center of the Landau level than determined above for the both N=0 and 1 levels.

So far, only the case of short-range scatterers has been considered. Figure 6 gives the density-of-states histogram of the N=0 Landau level for long-range ($d=2\ell$) scatterers. The corresponding histograms of g(L) and decay rate α of localized wave functions are shown in Figs. 7 and 8, respectively. Almost all the states except those which lie in the extreme vicinity ($|E|<0.05\Gamma_0$) of the center of the Landau level are localized for such long-range scatterers. As has already been discussed in section 1, the electron motion becomes nearly classical and electrons tend to be localized along an equipotential line when the range of the potential becomes much larger than the cyclotron radius. An example of the potential distribution is given in Fig.9 and wave functions corresponding to a state near the center of Landau level and a localized state are given in Fig.10. It is clearly demonstrated that the electron motion is confined along an equipotential region. The wave function of a state localized along a closed

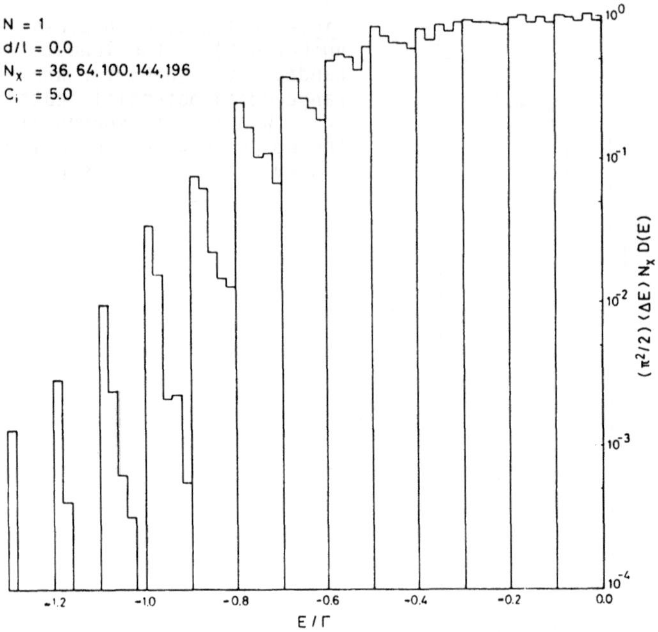

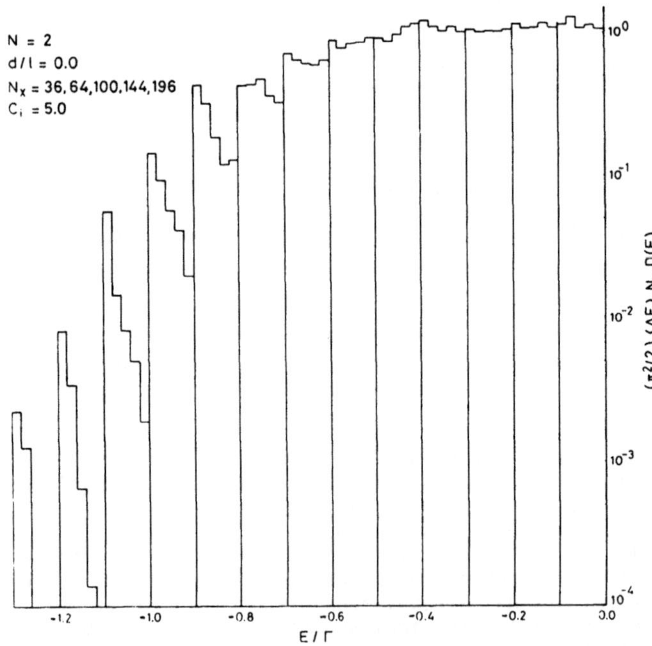

Fig.3 Calculated Thouless number g(L) of the first (N=1) and second (N=2) excited Landau levels for short-range delta-potential scatterers.

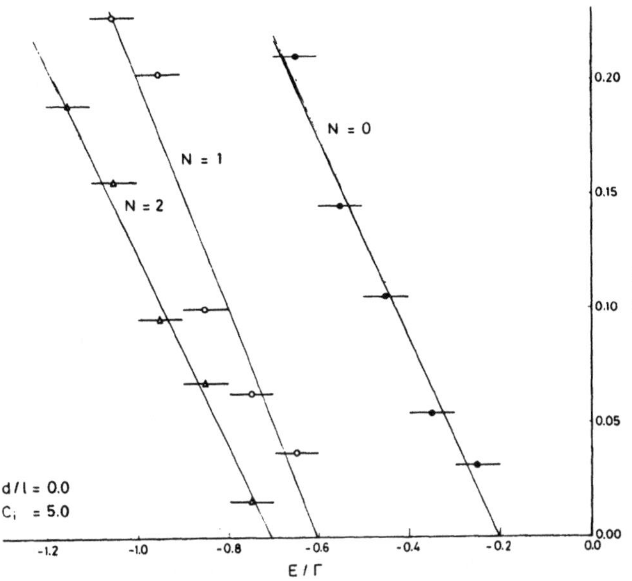

Fig.4 Calculated decay rate α of localized wave functions of the N = 0, 1, and 2 Landau levels as a function of energy. Short-range delta-potential scatterers are assumed.

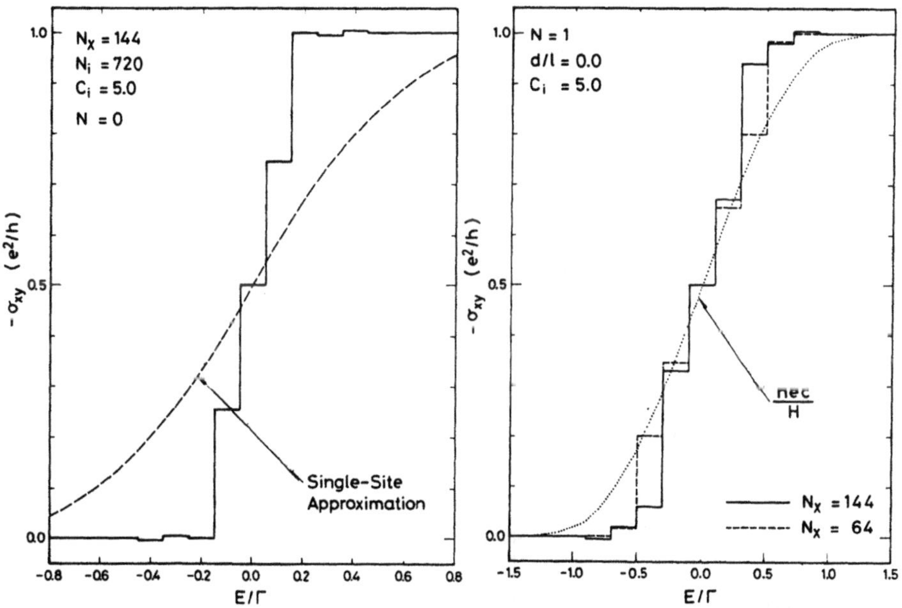

Fig.5 Calculated Hall conductivity of the N = 0 and 1 Landau levels as a function of the Fermi energy at the absolute zero of temperature. Short-range delta-potential scatterers are assumed. The result in the single-site approximation is given by the dashed line.

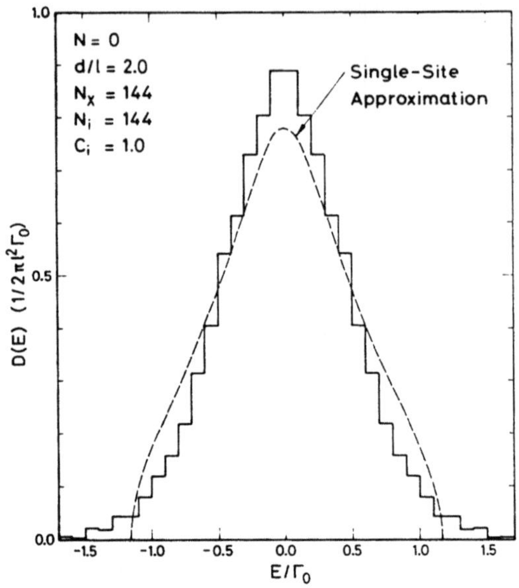

Fig.6 Calculated histogram of the density of states of the lowest Landau level for long-range scatterers (d=2ℓ). The result in a single-site approximation is given by the dashed line.

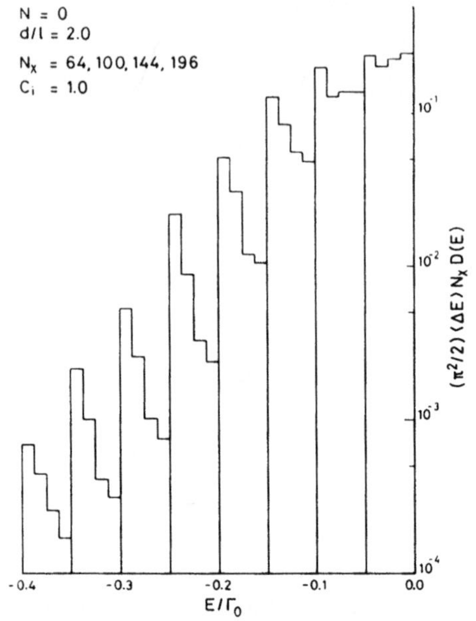

Fig.7 Calculated Thouless number g(L) of the lowest Landau level for long-range scatterers (d=2ℓ).

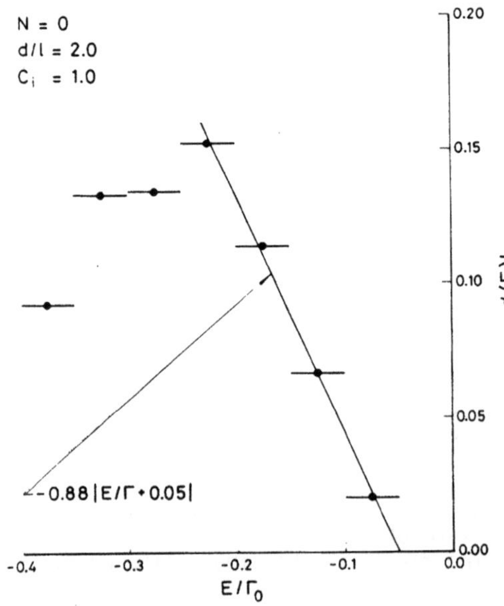

Fig.8 Calculated decay rate α of localized wave functions of the lowest Landau level for long-range scatterers (d=2ℓ).

Fig.9 An example of the potential distribution for long-range scatterers. The system size L is about 30 in units of ℓ. Positions of repulsive and attractive scatterers are denoted by + and -, respectively. The solid lines denote the equipotential lines for E=0.1, 0.3, 0.5, ... and the dashed lines for E=-0.1, -0.3, -0.5, ... in units of Γ_0.

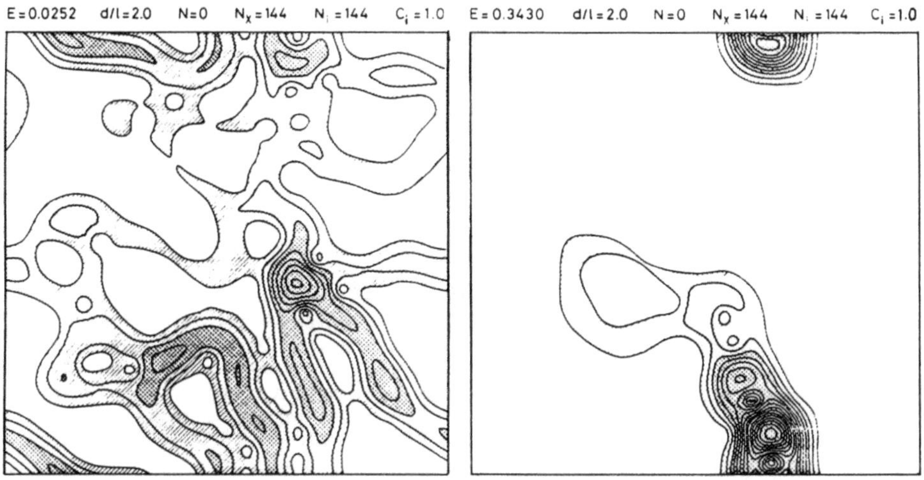

Fig.10 An example of wave functions for the potential given in Fig.9. The wave function of a level near the center of the Landau level and that of a localized state are depicted. The absolute value of the wave function is appreciable in the shaded region.

equipotential loop does not decay with distance exponentially, which might explain the fact that the L dependence shown in Fig.7 becomes weaker for large system size L. Figure 11 gives the corresponding Hall conductivity for N_X=196. The Hall conductivity starts to deviate from the integer multiple of $-e^2/h$ already below the "mobility edge" obtained from Fig.8. This is, however, considered to be more reasonable than for short-range scatterers since the system size used for the calculation is not infinitely large. It might also suggest that the localization is weaker and states can easily be delocalized by an electric field in case of long-range scatterers. Figure 9 shows that the simple estimate of the percolation threshold using a circular loop orbital [12] is quite inappropriate.

References

1. K. von Klitzing, G. Dorda and M. Pepper, Phys. Rev. Lett. 45, 494 (1980).
2. S. Kawaji and J. Wakabayashi, in Physics in High Magnetic Fields, ed. by S. Chikazumi and N. Miura, Springer Series in Solid-State Sciences, Vol. 21 (Springer, Berlin, Heidelberg, New York 1981) p. 284.
3. D.C. Tsui and A.C. Gossard, Appl. Phys. Lett. 38, 550 (1981).
4. T. Ando, Y. Matsumoto and Y. Uemura, J. Phys. Soc. Jpn. 39, 279 (1975); T. Ando, unpublished.
5. R.E. Prange, Phys. Rev. B23, 4802 (1981).
6. H. Aoki and T. Ando, Solid State Commun. 38, 1079 (1981).
7. R.B. Laughlin, Phys. Rev. B23, 5632 (1981).
8. D.J. Thouless, J. Phys. C 14, 3475 (1981).
9. S. Kawaji and J. Wakabayashi, Surf. Sci. 58, 238 (1976).
10. M. Tsukada, J. Phys. Soc. Jpn. 42, 391 (1977).
11. H. Aoki and H. Kamimura, Solid State Commun. 21, 45 (1977).
12. M. Tsukada, J. Phys. Soc. Jpn. 41, 1466 (1976).
13. H. Aoki, J. Phys. C 10, 2583 (1977); 11, 3823 (1978).
14. J.T. Edwards and D.J. Thouless, J. Phys. C 5, 807 (1972).
15. M. Kikuchi, J. Phys. Soc. Jpn. 37, 904 (1974).
16. D.C. Licciardello and D.J. Thouless, J. Phys. C 8, 4157 (1975).
17. R. Kubo, S.J. Miyake and N. Hashitsume, Solid State Phys. (ed. F. Seitz and D. Turnball), Vol. 17, Academic Press, New York (1965), p. 269.
18. N. Takemori, private communication.
19. T. Ando, J. Phys. Soc. Jpn. 36, 1521 (1974).
20. T. Ando and Y. Uemura, J. Phys. Soc. Jpn. 36, 959 (1974).
21. T. Ando, J. Phys. Soc. Jpn. 37, 622 (1974).
22. T. Ando, Y. Matsumoto, Y. Uemura, M. Kobayashi and K.F. Komatsubara, J. Phys. Soc. Jpn. 32, 859 (1972).

Localization in a Strong Magnetic Field

D.J. Thouless

Department of Physics, FM-15, University of Washington
Seattle WA 98195, USA

Abstract

Various aspects of the quantized Hall effect are discussed. Laughlin's argument for quantization of the Hall conductance in two dimensions is presented, and the parts played by localized states, extended states, and edge states are considered. Perturbative approaches are also considered, and it is shown that the Kubo formula gives a range in which there are no corrections to the quantized Hall current. The effect of a periodic substrate potential is discussed. The singular continuous spectrum produced by such a potential leads to unusual effects on conductivity. The Hall current must be entirely carried by one sub-band of the Landau band in this case.

1. Introduction

ANDO, MATSUMOTO and UEMURA [1] made an approximate calculation of the two-dimensional Hall effect some years ago, but the experimental discovery by VON KLITZING, DORDA and PEPPER [2] that the Hall conductance is quantized in multiples of $e^2/2\pi\hbar$ with an accuracy of at least one part in 10^6 was a startling one. It stimulated a lot of theoretical analysis which made it clear that the quantization should have been predicted. This work has already been discussed by ANDO [3]. The analysis is particularly important because it has been suggested that the effect can be used to improve the determination of fundamental constants. We therefore need to know the corrections to the quantized Hall conductance with a precision of one part in 10^8.

In this talk I repeat Laughlin's [4] analysis of the quantization condition. This analysis leads to several important conclusions. The quantization is exact if the Fermi level lies in a gap between Landau levels or in a region of localized states. There must be two-dimensional extended states, and there must be one-dimensional extended states at the edge of a sample, if the quantized Hall current is nonzero.

One can also get more insight into the problem by use of perturbation theory and the Kubo formula, and this is discussed in Sec. 3.

In Sec. 4 some recent work, done in collaboration with Dr. M. Kohmoto, on the effect of a periodic substrate potential in addition to the strong magnetic field is discussed. This may or may not be relevant to the problem of a random potential, but it is particularly interesting because it can lead to a singular continuous spectrum and so give a situation intermediate between localized and extended states. The periodic potential splits each Landau level into many sub-bands, and one can ask how the Hall current is split between these sub-bands.

Interactions between electrons are not discussed, since LEE [5] will consider interacting electrons in his talk.

2. Laughlin's Argument for Quantization of Hall Current

In this section I repeat an argument given originally by LAUGHLIN [4] in a form modified by HALPERIN [6]. Halperin considers an annulus of two-dimensional conductor bounded by an inner circle of radius R_1 and an outer one of radius R_2. The conductor is in a uniform magnetic induction B, and in addition a long solenoid passes through the center of the system, so that the flux, and so the vector potential, in the conductor can be changed without changing B. This system is shown in Fig.1. A radial electric field $\mathcal{E}$ can also be applied.

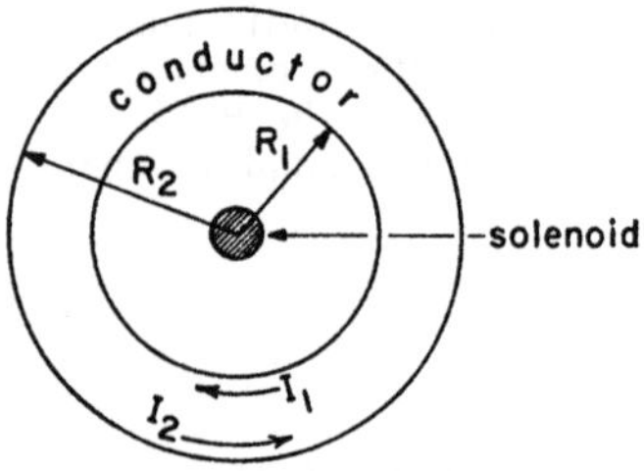

Fig.1 Configuration of the conductor used in the argument for quantized Hall current.

In the interior of a perfect conductor the wave functions have the appropriate form

$$\psi_{n,\nu}(r,\theta) \simeq \phi_n(r - \hbar\nu/eBr - m\mathcal{E}/eB^2)e^{i\nu\theta} \ , \tag{1}$$

provided the cyclotron radius is much less than r. The energy of this state is $(n+\frac{1}{2})\hbar eB/m$ if the electric field $\mathcal{E}$ is zero. The tangential current is given by

$$j = \frac{e}{m}\left(-i\frac{\hbar}{r}\frac{\partial}{\partial\theta} - eBr\right) = \frac{e}{m}\left(\frac{\hbar\nu}{r} - eBr\right) \ , \tag{2}$$

and this has the expectation value $-e\mathcal{E}/B$, the Hall current of a free electron. When ν has a value such that the center coordinate $\bar{r} = (\hbar\nu/eB)^{\frac{1}{2}}$ is within a cyclotron radius of or beyond one edge or the other, the energy is raised and there is a net current in the absence of an electric field which contributes to the Landau diamagnetism. If the electrochemical potentials at the two edges are equal the currents I_1 and I_2 cancel, but if there is a difference in this electrochemical potential (Fermi energy) there will be a resultant Hall current.

If the conductor is perfect close to the edges but disordered in the interior, so that the degeneracy of the Landau levels is broken, but they do not overlap, then a level scheme such as is shown in Fig.2 is obtained. The Fermi energy E_F is taken to lie between two Landau levels, so that only at

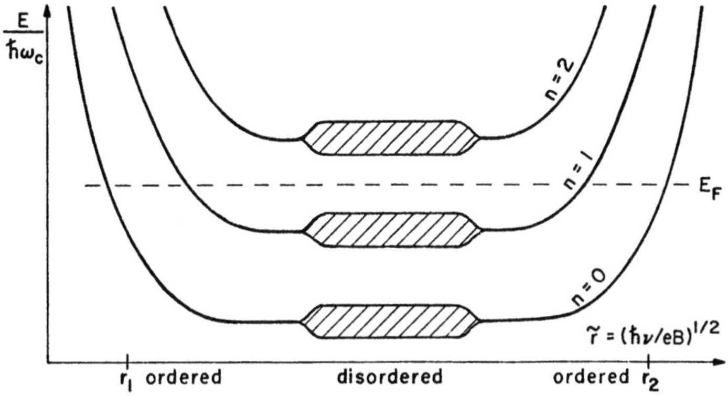

Fig.2 Energy levels as a function of position in the conductor shown in Fig.1

the edges are there states at E_F. Now the flux through the solenoid is changed continuously by an amount $2\pi\hbar/e$. As the flux is changed the eigenstates of the system change continuously since the vector potential is changing, but when the flux change is a multiple of $2\pi\hbar/e$ the eigenstates are identical except for a trivial phase factor. Thus a change of flux by $2\pi\hbar/e$ induces a mapping of the eigenstates onto themselves. The only change that can have occurred in the state of the system in the course of such a mapping is a transfer of electrons from just below the Fermi surface on one edge of the system to just above the Fermi surface on the other. In the case of a perfect conductor described by (1) the value of ν is changed by unity, and so one electron is transferred from one edge to the other for each fully occupied Landau level. For equilibrium the total change of energy must be zero when such a transfer is made. The energy change has an electrostatic contribution NeV_H, where N is the number of electrons transferred from one edge to the other and V_H is the potential difference between the edges, and a magnetic contribution

$$\int \underset{\sim}{j} \cdot \delta \underset{\sim}{A} d^2r = 2\pi\hbar J/e \ , \tag{3}$$

where $\delta \underset{\sim}{A}$ is the change of vector potential due to the $2\pi\hbar/e$ change of flux and J is the current. The equality of this to the electrostatic energy leads to the result

$$\frac{J}{V_H} = \frac{Ne^2}{2\pi\hbar} \ , \tag{4}$$

where N is some integer.

This argument can be looked at in a slightly different way. The change in flux generates an e.m.f. round the annulus while it is changing, and the electrons that are transferred from one edge to another constitute the Hall current that flows in response to this e.m.f.

There are several important implications of this argument. Localized states do not play any part in the response to a changing flux, since unless there is a significant amplitude of the state extending all round the annulus the changing flux simply induces a gauge transformation of the state. If there is a Hall current there must be extended states to respond to the changing flux, and in particular the edge states at the Fermi energy must be extended.

3. Perturbative Approaches

AOKI and ANDO [7], PRANGE [8], and THOULESS [9] have all considered the application of perturbation theory to the problem of two-dimensional electrons in a strong magnetic field. In the case of noninteracting electrons some rather strong statements about the convergence of perturbation theory and its consequences for the Hall current can be made.

Firstly, if the Fermi energy lies between Landau levels and the substrate potential is too weak to shift energies from the Landau levels to the Fermi energy, then perturbation theory converges. A precise statement of this is that ground state perturbation theory converges if

$$|V(\underset{\sim}{r})| < |E_F - (n+\tfrac{1}{2})\hbar\omega_c| \;, \quad \text{for all } \underset{\sim}{r} \text{ and all } n \;. \tag{5}$$

This establishes a finite radius of convergence for the perturbation series. Furthermore there is a domain within which the Green function falls off exponentially with distance [9], although this has not been proved over the whole domain given by (5).

The Kubo formula for the Hall conductivity can be written in the form

$$\sigma_{yx} = -\frac{e^2}{2\pi i A m}\oint \mathrm{tr}\left|\frac{1}{z-\hat{H}}\left(-i\hbar\frac{\partial}{\partial y} - eB\hat{x}\right)\frac{1}{z-\hat{H}}\,\hat{x}\right| dz \tag{6}$$

where A is the area of the system and the contour of integration surrounds the real axis below E_F. Provided the Green function falls off exponentially with distance at the Fermi energy (it certainly falls off exponentially everywhere else on the contour), the functional derivative of this expression with respect to the substrate potential can be taken and the relation between $\hat{j}_y$ and $[\hat{H},\hat{y}]$ used to get the result

$$\delta\sigma_{yx} = \frac{e^2}{2\pi\hbar A}\oint \mathrm{tr}\left(-\frac{1}{z-\hat{H}}\,\delta\hat{V}\,\frac{1}{z-\hat{H}}\,\hat{y}\hat{x} + \frac{1}{z-\hat{H}}\,\delta\hat{V}\hat{y}\,\frac{1}{z-\hat{H}}\,\hat{x} - \frac{1}{z-\hat{H}}\,\hat{y}\delta\hat{V}\,\frac{1}{z-\hat{H}}\,\hat{x} + \hat{y}\,\frac{1}{z-\hat{H}}\,\delta\hat{V}\,\frac{1}{z-\hat{H}}\,\hat{x}\right) dz \;. \tag{7}$$

If the support of $\delta\hat{V}$ is some finite region of space the integrals implied by the trace are all finite, since the Green function $(z-\hat{H})^{-1}$ falls off exponentially. The cyclic invariance of the trace makes this expression vanish, so there is no change of the Hall current induced by any change of the potential in a finite region, provided the Green function at the Fermi energy falls off exponentially.

These arguments show that there is a domain of convergence of perturbation theory, and within that domain all corrections to the free electron Hall conductivity vanish. Since the exponential decay of the Green function seems to be the key point of the argument it suggests that localized states at the Fermi energy will not change things. In this way LAUGHLIN's [4] results are derived by considering an infinite system rather than a bounded one with edges.

4. Periodic Perturbations in a Strong Field

In this section unpublished work carried out in collaboration with Dr. M. Kohmoto is described. It has been our hope that by studying the effect of a periodic potential we could learn something about the possible behavior in a random potential.

It was shown by HARPER [10] that if a potential of the form

$$V' \cos \alpha x + V \cos \alpha y \tag{8}$$

is treated perturbatively in the strong field limit a simple difference equation is obtained. If the Landau orbitals (1) are used as a basis, then the term $V' \cos \alpha x$ gives only a diagonal term in the secular matrix (the mixing of different Landau levels is neglected in the strong field limit), while the term $V \cos \alpha y$ changes the value of k by $\pm\alpha$. If the eigenvector of the secular matrix is written as

$$\psi = \sum_n c_n \phi[x - \hbar(n\alpha + k_0)/eB] \exp[i(n\alpha + k_0)y] \ , \tag{9}$$

then the amplitudes c_n satisfy the equation

$$Vc_{n-1} + 2V' \cos(\nu + 2\pi n/\phi)c_n + Vc_{n+1} = Ec_n \ , \tag{10}$$

where a factor irrelevant to this discussion has been removed from the energy E and

$$\phi = 2\pi eB/\hbar\alpha^2 \tag{11}$$

is the number of flux quanta per unit cell, while ν is a parameter that varies over a range $2\pi/\phi$. To lowest order in perturbation theory the components of the current operator can be represented by matrices whose nonvanishing elements are

$$(\hat{j}_x)_{n,n\pm 1} = \pm ieV/m \ , \quad (\hat{j}_y)_{nn} = (2eV'/m) \sin(\nu + 2\pi n/\phi) \ . \tag{12}$$

In the case of ϕ equal to a rational number p/q it is easy to show that each Landau level breaks up into p distinct bands, except that when p is even the two central bands touch at E = 0. AUBRY and ANDRE [11] have studied the case of ϕ irrational and shown that for $|V|<|V'|$ states are localized in the x direction and extended in the y direction. Numerical calculations show that the sum of the widths of the p bands is equal to $2|V - V'|$ in the limit of p infinite (ϕ irrational).

In this case localization does not imply that no current flows. In the case V' = 0 there is a sinusoidal potential in the y direction, so states are localized in the y direction. However, the effect of an electric field in the x direction is to move the electrons along the equipotentials of constant $V \cos \alpha y - \mathcal{E}ex$ with a y component of velocity $-\mathcal{E}/B$. This can be seen most easily by choosing the vector potential to be in the x direction, so that the wave function is centered at a value of y proportional to k_x. These Landau functions are eigenvectors of the secular matrix with energy $2V \cos(\alpha\hbar k_x/eB)$. The electric field increases k_x steadily at a rate $e\mathcal{E}/\hbar$, so y changes steadily at a rate $-\mathcal{E}/B$, while the velocity in the x direction is given by $(1/\hbar)dE/dk_x$, which is proportional to $(2V\alpha/eB) \sin(\alpha\hbar k_x/eB)$.

This anomalous behavior arises because there is a continuum of states localized in the y direction; in the usual case states close together in space must be separated in energy.

The case V = V', with square symmetry, is particularly interesting. The spectrum was studied in detail by HOFSTADTER [12], and he found a number of interesting properties. Firstly he found that if a sequence of rational values of $\phi = p/q$ are studied there is a scaling property, so that subsets of the bands for large p look like scaled-down versions of the bands for smaller p. This is illustrated in Fig.3, where the sequence of values for ϕ = 3, 5/2, 8/3, 13/5 is shown. In the irrational limit $p \to \infty$ the band widths all shrink to zero in such a way that the total width of the bands is zero, so that the spectrum is a singular continuous spectrum concentrated on a Cantor set of measure zero. In fact we have found that the total width of the bands tends rapidly to the form A/p, where A is a constant, independent of ϕ, when p is large.

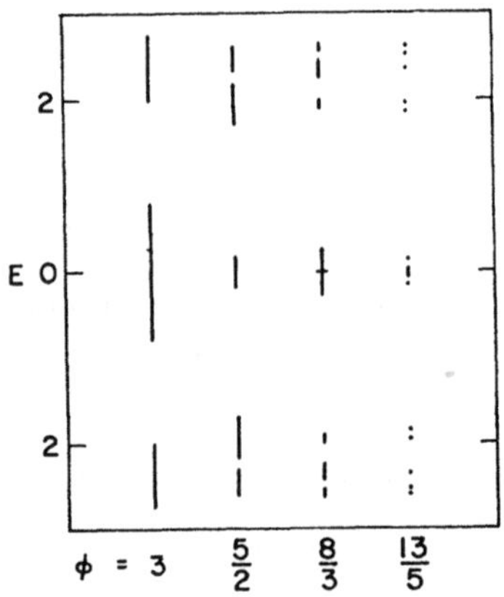

Fig.3 Sub-bands of a Landau level for various values of ϕ, the flux per unit cell

This singular continuous spectrum is intermediate between the absolutely continuous spectrum of extended states and the point spectrum of localized states. It obviously has important effects on the response to electromagnetic perturbations. If the Fermi energy is chosen arbitrarily it will in general lie in an energy gap--the probability that this gap is less than Δ is proportional to Δ^2--so that the system behaves as an insulator with a gap. If, however, a fixed number of electrons is taken, the Fermi energy generally lies arbitrarily close to regions of the spectrum. In this case one can calculate the behavior of $\int^{\omega}\sigma(\omega')d\omega'$ for small ω. I think it is quadratic in ω, but we have not yet completed this calculation. The integral is of course linear in ω when the Fermi energy lies in a region of extended states and cubic when it is in a region of localized states.

The behavior of the Hall current is particularly intriguing. LAUGHLIN's [4] argument shows unambiguously that the Hall current is quantized if the Fermi energy is in a gap. If ϕ is equal to 3 then the central sub-band carries all the current. Presumably for higher values of p the same is true, except that for even values of p it is shared between the two central bands that touch at E = 0. In the case of ϕ irrational the whole Hall current is carried by a vanishingly small fraction of the states at the center of the

Landau level. This is not quite as odd as it seems, since it is strictly true only for vanishingly small electric field. Any finite electric field will cause electrons to tunnel across the small gaps separating the sub-bands.

Acknowledgments

This work was supported in part by the National Science Foundation, grant no. DMR79-20785A01. I am grateful to Dr. M. Kohmoto and Dr. P. Nightingale for many useful discussions of this work.

References

1 T. Ando, Y. Matsumoto, Y. Uemura: J. Phys. Soc. Japan 39, 281 (1975)
2 K. von Klitzing, G. Dorda, M. Pepper: Phys. Rev. Lett. 45, 494 (1980)
3 T. Ando, this Symposium
4 R.B. Laughlin: Phys. Rev. B 23, 5632 (1981)
5 P.A. Lee, this Symposium
6 B.I. Halperin, preprint
7 H. Aoki, T. Ando: Solid State Commun. 38, 1079 (1981)
8 R.E. Prange: Phys. Rev. B 23, 4802 (1981)
9 D.J. Thouless: J. Phys. C 14, 3475 (1981)
10 P.G. Harper: Proc. Phys. Soc. Lond. A 68, 874 (1955)
11 S. Aubry, G. André: in Non-linear Structure and Dynamics in Condensed Matter (Springer-Verlag, Berlin, 1978), p. 131
12 D.R. Hofstadter: Phys. Rev. 114, 2239 (1976)

The Quantized Hall Effect and Other Macroscopic Quantum Phenomena

Y. Imry[1]

IBM Thomas J. Watson Research Center
Yorktown Heigths, NY 10598, USA

The relation between the quantized Hall conductance and other macroscopic quantum effects is discussed. Under the appropriate conditions, fluxoid quantization and a phenomenon similar to the AC Josephson effect are possible in normal conductors.

1. Introduction

The recent observation [1-3] of the quantized Hall conductance [4,5] constitutes a remarkable quantum mechanical effect on a macroscopic scale. Apart from diamagnetism and quantum oscillations in metals, such macroscopic quantum phenomena have always been associated with the superconducting state. In fact, some explanations [6,7] of e.g. flux quantization and the AC Josephson effect, while having been applied to the case of superconductivity, were relying on much more general principles (e.g., gauge invariance). The superconductivity was in fact only necessary to provide a macroscopic amplitude for the circulating currents necessary to establish these phenomena, and to make the charge q appearing in the flux quantum hc/q become q=2e. Some time ago, it has been argued [8] that these effects are also possible in cases where the superconducting order parameter (ODLRO) [9] vanishes due to fluctuations in a thin superconductor, as well as in normal metal rings under conditions of good purity and low temperatures. The latter conditions are in principle feasible, but not too easy to achieve in practice. We shall find in this paper that systems which exhibit quantized Hall conductance should theoretically also show the above quantum phenomena. This would mean that such macroscopic quantum effects, with a flux quantum hc/e and Josephson frequency $eV/\hbar$, are indeed possible, in principle, in normal systems.

Our discussion will be in terms of Laughlin's [10], (see also [11]) extremely instructive explanation of the quantized Hall effect. Laughlin's ring geometry is reminiscent of the Byers and Yang [6] multiply-connected geometry. In fact, by combining these two analyses we arrive at the claimed conclusions.

---

[1]On leave of absence from the Department of Physics and Astronomy, Tel-Aviv University, Tel-Aviv 69978, Israel.

We emphasize that the similarity of the quantized Hall current to a supercurrent was already mentioned in Laughlin's paper. More recently, AZBEL [12] has termed this an "ideal conductor", and suggested that many weak insulators may display such an ideal conductivity. AZBEL'S work has provided some of the stimulation for this one. Here we consider specifically the two phenomena of flux quantization and the AC Josephson effect, in a system without ODLRO.

2. General Theorems

We start by applying the BYERS-YANG and BLOCH results, originally obtained in the general geometry of Fig. 1a, to the simplest Hall-effect geometry of LAUGHLIN - Fig. 1b.

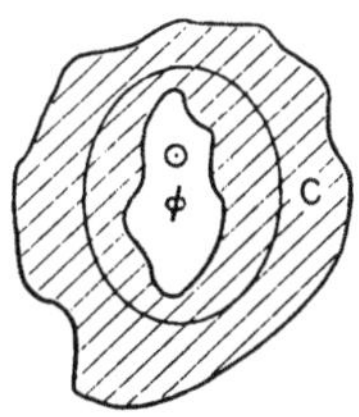

Fig. 1a The general multiply connected geometry of BYERS and YANG - a magnetic flux Φ pierces through the hole in the material. The curve C goes around the hole.

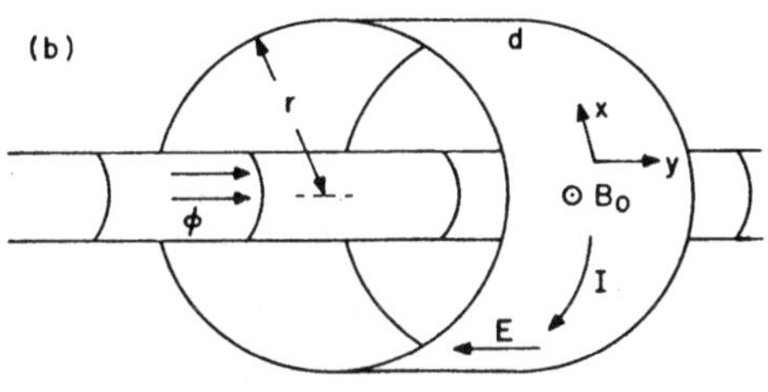

Fig. 1b LAUGHLIN'S geometry for the quantized Hall effect. The thin film is wrapped on a cylinder of radius r and height d. A radial magnetic field $\mathbf{B}_0$ and an electric field E parallel to the cylinder axis are applied. The current flows as shown. The gauge flux Φ is confined to an inner cylinder inside the hollow in the larger one.

The general periodicity of the physical properties of the system with the enclosed flux in the ring holds also in the presence of the Hall magnetic and electric fields. Since it is also insensitive to static fields, disorder, and electron-electron interactions, this periodicity is very general.

We consider a thin film wrapped on the outer cylinder of Fig. 1b. The electrons, whose coordinates are denoted by $\mathbf{r}_j$, are subject to a strong magnetic field $\mathbf{B}_0$ perpendicular to the film, to an electric field $\mathbf{E}$ parallel to the axis of the cylinder and to electron-ion and electron-electron interactions. The three latter effects are represented, taking the ions to be static, by a potential energy $V(\mathbf{r}_j)$ and the magnetic field $\mathbf{B}_0$ is represented by a vector potential $\mathbf{A}_0$ such that $\mathbf{B}_0 = \nabla \times \mathbf{A}_0$.

In addition, we also envisage a "gauge" AHARONOV-BOHM type [13] magnetic field leading to a flux Φ through the inner opening of the cylinder. For definiteness this flux may be thought of as going totally through a smaller-radius cylinder (e.g. a part of a toroid closing itself further away) piercing the inside of the larger one. The magnetic field $\mathbf{B}_g$ associated with this flux is strictly zero in the material. It is represented by a gauge vector potential $\mathbf{A}_g$ such that $\int_C \mathbf{A}_g \cdot d\mathbf{l} = \Phi$ where C is any closed curve in the $\mathbf{B}_g = 0$ region enclosing the flux Φ . C will be taken to be inside the material, where $\mathbf{A} = \nabla \chi$ and χ changes by Φ during each turn around the cylinder.

The Hamiltonian for the electrons is

$$\mathcal{H} = \frac{1}{2m} \sum_j \left[\frac{\hbar}{i} \nabla_j + \frac{e}{c}(\mathbf{A}_0(\mathbf{r}_j) + \mathbf{A}_g(\mathbf{r}_i))\right]^2 + V(\mathbf{r}_1, ..., \mathbf{r}_N) \tag{1}$$

Clearly, multiplying the wave function by the phase factor

$$\exp\left[-i \sum_j \frac{e}{\hbar c} \chi(\mathbf{r}_j) \right] \tag{2}$$

eliminates $\mathbf{A}_g$, and hence the flux Φ , from the equation, except that the boundary condition is now that the phase of the wave function changes by

$$\Delta\phi \equiv \frac{e}{\hbar c} \Phi = 2\pi\Phi/\Phi_0; \ \ \Phi_0 = hc/e \tag{3}$$

when any single electronic coordinate is taken once around the ring. This boundary condition is invariant under a change of Φ by an integer multiple of Φ_0. Thus, the BYERS-YANG theorems are established for the Hall effect case as well. The energy levels are periodic functions of Φ with period Φ_0, as are the partition function, the free energy F, and hence the total circulating current

$$I = -c \frac{\partial F}{\partial \Phi} \xrightarrow[T \to 0]{} -c \frac{\partial E_0}{\partial \Phi} \tag{4}$$

where E_0 is the ground state energy. Note, however, that the even (odd) character of the energy (current) as function of Φ is lost, in general, in the presence of the field $\mathbf{B}_0$.

It is instructive to remind ourselves how (4) is derived. The effect of an infinitesimal $d\phi$ is to cause a corresponding change of $\mathbf{A}_g$ in (1). This change is now regarded as a perturbation to the Hamiltonian, and the first-order change it produces on an energy level is calculated. It turns out that this change is equal to the average of the current operator, in the appropriate eigenstate, times $d\phi$.

Thus, (4) is valid provided the system stays in the same state (ground state at T=0, or in thermodynamic equilibrium, for T≠0) during the change in the gauge flux ϕ and the effect of $d\phi$ can be treated by perturbation theory. Thus, the change of ϕ should be *slow* enough on the scale of the relevant response times of the system. We shall return to this condition later. It should also be emphasized that the current as given by (4) is an equilibrium (or ground-state) property. No dissipation is associated with it, and it is therefore of a different nature from a regular transport current.

It thus follows immediately that the energy can *not* be a monotonic function of Φ leading to a constant Hall current. It should rather be like in Fig. 2a, exhibiting the required periodicity. This yields an $I(\Phi)$ like in Fig. 2b which implies fluxoid quantization and/or the AC Josephson effect, using the arguments of BYERS and YANG and BLOCH as discussed below. In order to understand how the periodic structure comes about, it is useful to consider the simple case of free electrons, combining the anaylses of LAUGHLIN [10] and HALPERIN [11]; this is done in the next section.

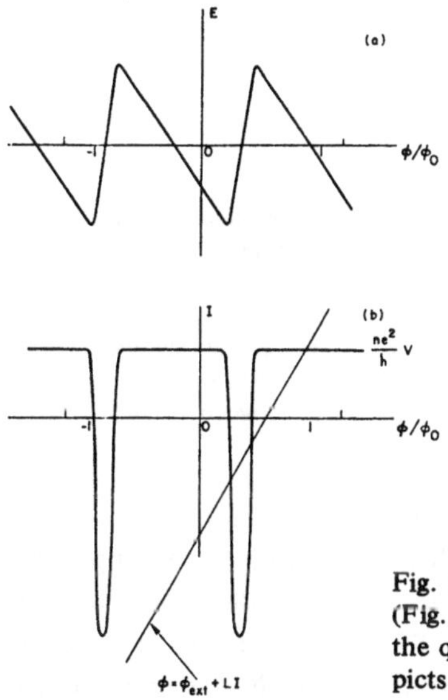

Fig. 2 A schematic representation of $E(\Phi)$ (Fig. 2a) and $I(\Phi)$ (Fig. 2b) for the geometry of Fig. 1b under conditions where the quantized Hall effects exists. The straight line in Fig. 1b depicts the geometrical relation (7) between Φ, Φ_{ext} and I.

3. The Case of Free Electrons

We employ the simpler ribbon geometry due to Laughlin and his local x-y coordinate system around the cylinder. $\mathbf{A}_0$ is given in the Laudau gauge by $\mathbf{A}_0 = (B_0y,0,0)$, while the gauge part is $\mathbf{A}_g = (\Phi_0/2\pi r,0,0,)$. The eigenstates are plane waves in x, e^{ikx}, with $k = (2\pi/L)$ times an integer, and Landau harmonic oscillator levels in the y direction with the cyclotron frequency $\omega_c = eB_0/mc$. The harmonic oscillator for a given k and Φ is centered around

$$y_0 = \hbar k/m\omega_c + eE_0/m\omega_c^2 - \Phi/2\pi rB_0 \tag{5}$$

and the height of its bottom is eEy_0 + const. Due to the quantization of k, the allowed y_0 values (see Fig. 3) are separated by $\Delta y_0 = \hbar/m\omega_c r$. For a given Landau level number i the energy level depends on y_0 through eE_0y_0 and in addition levels whose y_0 is within a cyclotron radius $\ell_H \equiv (\hbar c/eB_0)^{1/2}$ of the sample edges (y = 0,d) will feel the boundary condition there. This important effect (note that usually $\ell_H >> \Delta y_0$) was introduced by HALPERIN for this problem. The total dependence of a given energy level on its y_0 is schematically depicted in Fig. 3.

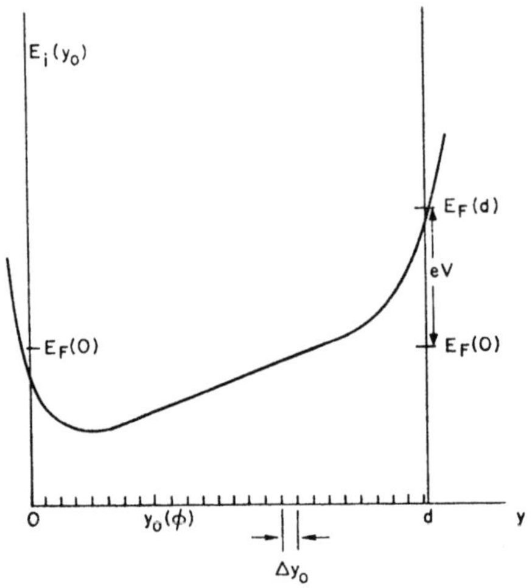

Fig. 3 A schematic dependence of the energy of a given Landau level on the position of the minimum, y_0, for V = Ed. The levels are filled such that $E_F(d) - E_F(0) = eV$.
The markings on the y axis signify the allowed values of y_0 (Eq. 5) for a given value of ϕ, and the distance between them are Δy_0.

For a given number of free electrons with a potential difference across the film $V = Ed = [E_F(d) - E_F(0)]/e$, the electrons fill the levels up to the appropriate Fermi energies $E_F(d), E_F(0)$ at the two sides. We consider the situation appropriate to the quantized Hall effect where the Fermi level is always between the nth (full) and $(n+1)^{st}$ (empty) Landau level. When Φ is decreased by $\delta\Phi << \Phi_0$ all the y_0's will decrease by $\delta\Phi/2\pi rB_0$ (see (5)). Neglecting, for the moment, the surface effects, this will decrease the total energy by $Ende\delta\Phi/2\pi\Delta y_0 rB_0$. This is seen from (4) to yield the correct quantum Hall current ne^2V/h (spin and valley degeneracy have to be included in a more realistic treatment). However, when $\delta\Phi$ is increased further, then, at some stage, it will be necessary to depopulate one state at $y \cong 0$ and fill the one at $y \cong d$. This is the pumping of electrons discussed by LAUGHLIN. It occurs, assuming the system stays in the ground state, once in each cycle of increasing $\delta\Phi$ by Φ_0 and is easily seen to *exactly compensate* for the decrease in E due to the level lowering during the rest of the cycle. Halperin's analysis shows that the latter is unchanged when the surface effects are taken into account and opposite surface currents flow at y = 0 and y = d. In fact, the energy levels and their populations are invariant to changes of Φ by Φ_0, as required. Thus, we are led to the periodic $E(\Phi)$ curve depicted in Fig. 2a. The periodicity is due to the *combined* effects of energy-level lowering through most of the cycle and "pumping" across the potential through a small part of it. The small width of the latter process (necessary to establish (4)), for which we do not have a full picture, except in the narrow film case discussed below, may be due to disorder broadening of the levels and/or a smearing by a finite temperature. These effects are thought to be small under the conditions $k_BT<<\hbar\omega_c,V$ and $\hbar/\tau_{el}<<\hbar\omega_c,V$. When the temperature is increased and/or the magnetic field and/or purity are decreased, the structure is gradually smeared and should wash out completely (normal case) if at least one of the above conditions is not met. As emphasized by LAUGHLIN and earlier by BYERS and YANG and BLOCH, in a different context, these results follow from *general principles* and are only illustrated by the free electron case.

In the simple case, where the film is very narrow in the y direction ($d \gtrsim \Delta y_o$) only one or a few y_o values will be relevant. The eigenstates will be some linear combinations of these harmonic oscillator states, with coefficients that will be periodic in ϕ and the sharp structure in Fig. 2 will not exist. Eq. (4) will be more obviously valid and the time constants of the system may depend on the eigenstate mixing. This mixing is exponentially small for large d. Note that in the above small d case the amplitude of $I(\phi)$ will be much smaller than ne^2V/h and it will vanish when $d/\Delta y_0 \to 0$.

4. Macroscopic Quantum Phenomena

Until now, we observed that once the quantized Hall current is observed, the periodicity in Φ is a general, exact theorem. We now remind the reader how the phenomena of fluxoid quantization [6] and AC Josephson effect [7] follow from the periodic $I(\Phi)$. The construction due to BLOCH (7) is most useful here. For a given external flux Φ_{ext} the total flux is given by (see Fig. 2b.)

$$\Phi = \Phi_{ext} + LI \qquad (6)$$

where L is the self-inductance of the system. The physical states of the system are given by the intersections of (6) with the periodic $I(\Phi)$, which are minima of E (or of F at finite temperatures). Two cases are possible: a) When the straight line (6) is very flat, there are many intersections representing quantized (with Φ_0 of a single change e) metastable fluxoid states of the ring (the most stable being in general the one with a smallest shielding current). b) When the line (7) is very steep, one has only one intersection which is a stable state of the system. A dc voltage in the ring can only be simulated by a dc induced EMF, namely, $\Phi_{ext} = cV_1t + \text{const}$. This moves the straight line to the right at a rate cV_1 and yields a periodic current with a fundamental frequency

$$\omega = eV_1/\hbar \tag{7}$$

i.e., *the Josephson relation with a charge e* in this case. Note also that a periodic current with the same period will flow in the y direction. This current, for the case of Fig. 1b, will consist of bursts of current, each having a total charge of one electron (up to the appropriate degeneracies). If the ring is opened and a voltage V_1 is impressed paralled to I, one may argue, but this is less rigorous than the foregoing discussion, that the system cannot physically distinguish between an induced and a direct EMF - hence the AC Josephson effect. This is the line of reasoning implied by BLOCH [7]. In the superconducting case, the current amplitude, to achieve case b, may be reduced by introducing a tunnel barrier. In our case, this can be done, as discussed above, also by reducing B_0 or increasing T. The crucial point is to achieve a *non-zero amplitude* of $I(\Phi)$. This is the real distinction between systems that do and do not show these interesting phenomena. Of course, in an experimental demonstration of this effect, elimination of noise may also play an important role.

A crucial condition for the applicability of these effects is that the system should stay in the ground state or in thermodynamic equilibrium during the increase of Φ. The time it takes to change Φ by a quantity on the order of Φ_0 should be longer than, for example, the appropriate RC and L/R time constants of the system. Since the resistance of the system in the quantized Hall current state is very high, the RC time constant can be reduced by applying the Hall voltage V with a voltage source and measuring the resulting current I. This source will have a very small resistance, R_s to enable the electrons to be transferred across it at a high enough rate. (However, R_s must not be too small to make the L/R time short enough as well). The author is indebted to D. J. THOULESS for a discussion on this point during the symposium.

5. Delocalization in Strong Fields

An outstanding question here is [11,14,15,16] how are the extended states restored by the large B_0, to carry the current and to be sensitive to ϕ. For $B_0 = 0$, it is thought that all states are localized [17]. We do not discuss this question fully here, except to note [18] that there is an interesting connection between the $E(\Phi)$ dependence and the THOULESS criterion for localization [19,20]. In fact, the gauge transformation, discussed in section 2, shows that a change in Φ by $\Phi_0/2$ is equivalent to a change (e.g., from periodic to antiperiodic) in the boundary condition on the gauge-transformed wave function. Thus

$\Delta \sim E(\Phi + \Phi_0/2) - E(\Phi)$ yields the sensitivity of the energy level to boundary conditions. Multiplying this by the density of states per unit energy yields the THOULESS number [17] which determines whether the state is localized or not. Obviously, the gauge transformation does not alter the localization property of the state. Since y_o is changed by $\Delta y_0/2$ during the above process, one can readily obtain Δ from the change in eEy_0. If the effective E is large enough (as happens near the surface [11] or in the bulk, due to a large Hall voltage) it turns out that the state will not be strongly localized [18] once $\omega_c \tau >> 1$, even when L is larger than the localization length for $B_0 = 0$. The additional delocalization in a large **E** is in agreement with results reported by Ando [21] in this symposium. It is important to emphasize that the delocalization of the surface states, which appear one-dimensional, is due to motions *perpendicular* to the surface and hence is not a one-dimensional effect.

6. Summary

To summarize, we combined the LAUGHLIN analysis of the quantum Hall effect with the BYERS-YANG and BLOCH pictures for macroscopic quantum phenomena that have been associated with superconducting properties (see, however, Ref. [8]). We find that a system which shows the quantum Hall effect may also display other effects such as fluxoid quantization and Josephson type phenomena. These systems are in a sense insulators ($\sigma_{xx} = 0$). Azbel (1981) has recently argued that weak insulators will, more generally, display *ideal conductivity* under conditions less stringent (smaller B_0, higher T) than those needed for the quantized Hall effect. This raises the interesting possibility that the phenomena considered here may have a much more general range of applicability. We also presented arguments, using the above picture, for the delocalizing effect of a strong magnetic (as well as a perpendicular electric) field.

Acknowledgements

The author thanks L. GUNTHER for many past discussions on related subjects and D. J. THOULESS, A. B. FOWLER, S. KIRKPATRICK, R. W. LANDAUER and F. STERN for more recent discussions. He also thanks M. YA AZBEL and B. I. HALPERIN for preprints of their papers.

References

1 K. Von Klitzing, G. Dorda, and M. Pepper, Phys. Rev. Lett. **45**, 494 (1980).
2 D. C. Tsui and A. C. Gossard, Appl. Phys. Lett. **38**, 550 (1981).
3 S. Narita, S. Takeyama, W. B. Luo, S. Hiyamizu, K. Nanbu, and H. Hashimoto, preprint (1981).

[4] T. Ando, J. Phys. Soc. Japan **37**, 622 (1974).
[5] T. Ando, Y. Matsumoto, and Y. Uemura, J. Phys. Soc. Japan **39**, 279 (1975).
[6] N. Byers and C. N. Yang, Phys. Rev. Lett. **7**, 46 (1961).
[7] F. Bloch, Phys. Rev. Lett. **21**, 1241; Phys. Rev. **166**, 415 (1968).
[8] L. Gunther and Y. Imry, Solid State Comm. **7**, 1391 (1969).
[9] C. N. Yang, Revs. Mod. Phys. **34**, 694 (1962).
[10] R. B. Laughlin, Phys. Rev. B **23**, 5632 (1981).
[11] B. I. Halperin, Harvard preprint (1981).
[12] M. Ya Azbel, University of Pennsylvania preprint (1981).
[13] Y. Aharonov, and D. Bohm, Phys. Rev. **115**, 485 (1959).
[14] R. E. Prange, Phys. Rev. B **23**, 4802 (1981).
[15] D. J. Thouless, J. Phys. C **14**, 3475 (1981).
[16] D. C. Tsui and S. J. Allen, Jr., Phys. Rev. B **24**, 4082 (1981).
[17] E.Abrahams, P. W. Anderson, D. C. Licciardello, and T. V. Ramakrishnan, Phys Rev. Lett. **42**, 673 (1979).
[18] Y. Imry and S. Kirkpatrick, to be published.
[19] D. J. Thouless, Phys. Rev. Lett. **39**, 1167 (1977).
[20] D. C. Licciardello, and D. J. Thouless, J. Phys. C **8**, 4157 (1975); ibid. **11**, 925 (1978).
[21] T. Ando, these proceedings (1981).

A Calculation of Two-Dimensional Hall Conductivity Under Strong Magnetic Fields – Wigner Representation

Y. Ono

Department of Physics, University of Tokyo
Bunkyo-ku, Hongo 7-3-1, Tokyo 113, Japan

Abstract

By noting the commutation relation $[X,Y] = i\ell^2$, where X and Y are the two components of the center coordinate for the cyclotron motion of a two dimensional electron in a perpendicular magnetic field H and $\ell = \sqrt{c\hbar/eH}$ is the cyclotron radius, it is argued that the Wigner representation with ℓ^2 replacing $\hbar$ in the usual case is very convenient in calculating the conductivity tensor under strong magnetic fields. This representation gives us an expansion parameter, which is the ratio of ℓ to the characteristic length describing the space variation of the scattering potential. In the very strong field limit, the Hall conductivity is expected to show a step-like behaviour as a function of the Fermi energy.

1. Introduction

Recently the localization problem in two dimensional disordered systems has drawn a great deal of attention. Especially, Abrahams, Anderson, Licciardello and Ramakrishnan [1] combined the perturbational calculation of the conductivity from the metallic side and the renormalization group approach in terms of the Thouless number [2], and derived the conclusion that particles moving in a two dimensional random potential are always localized irrespective of the strength of the potential fluctuation as far as it is finite. Vollhardt and Wölfle [3] extended the perturbational calculation and developed diagramatic self-consistent theory for the diffusion constant. Their self-consistent scheme gives the result consistent with that of Abrahams *et al.* The diagramatic self-consistent theory was possible due to the existence of a smallness parameter $1/\varepsilon_F\tau$ where ε_F is the Fermi energy and τ is the elastic relaxation time. Similar calculation can be performed even in the presence of a magnetic field if the field is weak enough, $\omega_c = eH/mc \ll \tau^{-1}$, ε_F, and it has been shown that a weak field does not break the localization in two dimensions [4,5].

In the case of strong magnetic fields, however, the situation is much different from that in the case of vanishing or weak fields. A strong magnetic field quantizes the electronic states, and if there is no potential, the density of states becomes a series of δ-functions located at $E = \hbar\omega_c(N + 1/2)$ with $N = 0,1,2,\cdots$ (the Landau levels). The scatterings due to impurities

In the following we assume that the cyclotron frequency is much larger than the subband width and that the inter-subband transitions can be neglected. Then, we have only to take N' = N in eq.(2.9), and the operator $\tilde{V}_N \equiv \tilde{V}_{NN}$ is written in the form,

$$\tilde{V}_N(X,Y) = (2\pi)^{-2}\iint dq\, V_q e^{-q^2\ell^2/4} L_N\left(\frac{q^2\ell^2}{2}\right) e^{i(q_xX+q_yY)} \quad (2.10)$$

where V_q is the Fourier transform of V(r), $L_N(z)$ the N-th order Laguerre polynomial. Note that $Y = (\ell^2/i)\partial/\partial X$ has the meaning of the y-coordinate of the center and that X and Y do not commute with each other. From the arguments given above, we find $\tilde{V}_N$ is nothing but the effective Hamiltonian for the center coordinate.

3. Wigner Representation

Hereafter we treat the following Hamiltonian,

$$\hat{\mathcal{H}} = (2\pi)^{-2}\iint dq\, v(q)\exp[i(q_xX+q_yY)]. \quad (3.1)$$

The information about the subband index and the impurity potential is included in v(q). In the following, we assume for the sake of simplicity that the Fermi energy ε_F lies within the lowest Landau subband N = 0, and neglect the contributions from higher subbands. Substituting the above Hamiltonian into eqs.(2.1) and (2.2), we can calculate the conductivities. The dynamical variables involved in these expressions are only X and Y. Although the method described below is applicable also to the calculation of the longitudinal conductivity, we will concentrate ourselves on the Hall conductivity in this work in order to avoid too much complexity.

When we consider the strong field limit where ℓ tends to zero, it is convenient to use the Wigner representation of the physical quantities such as

$$X_W = X - \frac{\ell^2}{2i}\frac{\partial}{\partial Y}, \quad (3.2)$$

$$Y_W = Y + \frac{\ell^2}{2i}\frac{\partial}{\partial X}, \quad (3.3)$$

$$\mathcal{H}_W = (2\pi)^{-2}\iint dq\, v(q)\exp[i(q_xX_W+q_yY_W)] . \quad (3.4)$$

Using the general theory of the Wigner representation [12], $\delta\sigma_{xy}$ (= σ_{xy} + nec/H) can be written in the form,

$$\delta\sigma_{xy} = \frac{e^2}{V}\int_0^\infty dt\int_0^\beta d\lambda\iint\frac{dXdY}{2\pi\ell^2}\,\{f(\mathcal{H}_W)\dot{Y}_W[1-f(\mathcal{H}_W)]\dot{X}_W(t+i\hbar\lambda)\}\cdot 1 , \quad (3.5)$$

and similarly eq.(2.4) is rewritten as

$$n = \frac{1}{V}\iint\frac{dXdY}{2\pi\ell^2}\, f(\mathcal{H}_W)\cdot 1 . \quad (3.6)$$

In eq.(3.5),

$$\dot{Y}_w = \frac{i}{\hbar}[\mathcal{H}_w, Y_w] \ ; \quad \dot{X}_w = \frac{i}{\hbar}[\mathcal{H}_w, X_w] \ , \tag{3.7}$$

and

$$\dot{X}_w(\tau) = e^{i\tau L}\dot{X}_w \tag{3.8}$$

with

$$iL = \frac{i}{\hbar}(\mathcal{H}_w - \mathcal{H}_{\bar{w}}) \ , \tag{3.9}$$

where $\mathcal{H}_{\bar{w}}$ is the same as $\mathcal{H}_w$ except for the replacement of ℓ^2 by $-\ell^2$ in X_w and Y_w.

If we note that

$$\exp[i(q_x X_w + q_y Y_w)] = \exp[i(q_x X + q_y Y)]\exp[-\frac{\ell^2}{2}(q_x\frac{\partial}{\partial Y} - q_y\frac{\partial}{\partial X})] , \tag{3.10}$$

the expansion in powers of ℓ^2 can be excuted straightforwardly. In the lowest order in ℓ^2, we have

$$\mathcal{H}_w = U(X,Y) + O(\ell^2) \ , \tag{3.11}$$

$$iL = -\frac{\ell^2}{\hbar}(\frac{\partial U}{\partial X}\frac{\partial}{\partial Y} - \frac{\partial U}{\partial Y}\frac{\partial}{\partial X}) + O(\ell^6) \ , \tag{3.12}$$

$$\dot{Y}_w = -\frac{\ell^2}{\hbar}\frac{\partial U}{\partial X} + O(\ell^4) \ , \tag{3.13}$$

$$\dot{X}_w = \frac{\ell^2}{\hbar}\frac{\partial U}{\partial Y} + O(\ell^4) \ , \tag{3.14}$$

where

$$U(X,Y) = (2\pi)^{-2}\int\int dq \ v(q)\exp[i(q_x X + q_y Y)] \ . \tag{3.15}$$

These expansions clearly show that, in the strong field limit ($\ell^2 \to 0$), X and Y move along the locus with constant U(X,Y) as is well known. Substituting eqs.(3.11)-(3.14) into eq.(3.5), we obtain

$$\delta\sigma_{xy} = -\frac{e^2}{V}\int_0^\infty dt\int_0^\beta d\lambda\int\int\frac{dXdY}{2\pi\ell^2}\frac{\ell^2}{\hbar}\frac{\partial U}{\partial X} f(U)[1-f(U)]e^{-\hbar\lambda L}e^{itL}\frac{\ell^2}{\hbar}\frac{\partial U}{\partial Y} \ . \tag{3.16}$$

The following relation,

$$e^{itL}\frac{\ell^2}{\hbar}\frac{\partial U}{\partial Y} = \frac{d}{dt}(e^{itL}X) \ , \tag{3.17}$$

allows us to perform the time integral in eq.(3.16) and the result can be written schematically as

$$\delta\sigma_{xy} = e^2\{\langle\dot{Y}(0)X(\infty)\rangle - \langle\dot{Y}(0)X(0)\rangle\} \ . \tag{3.18}$$

The first term in the curly bracket will be decoupled as $\langle\dot{Y}\rangle\langle X\rangle$ if the system is ergodic; we assume the ergodicity implicitly.

The average $\langle X\rangle$ is of the order of the system size but $\langle\dot{Y}\rangle$ will vanish because of the symmetry. Therefore the first term in the curly bracket in eq.(3.18) may be neglected. Thus $\delta\sigma_{xy}$ is rewritten as follows,

$$\delta\sigma_{xy} = \frac{e^2}{V}\frac{\ell^2}{\hbar}\iint\frac{dXdY}{2\pi\ell^2}\,\beta\,\frac{\partial U}{\partial X}\,f(U)[1-f(U)]X \ , \tag{3.19}$$

in the lowest order in ℓ^2. The Hall conductivity is now expressed as

$$\sigma_{xy} = -\,\frac{nec}{H} - \frac{e^2}{Vh}\iint dXdY\,\frac{\partial f(U)}{\partial X}\,X \quad . \tag{3.20}$$

If the sample has a rectangular form and covers the region, $X_0 < X < X_1$ and $Y_0 < Y < Y_1$, then a partial integration reduces the above expression into

$$\sigma_{xy} = -\frac{e^2}{h}\frac{1}{L_xL_y}\int_{Y_0}^{Y_1}dY\ [X_1f(U(X_1,Y)) - X_0f(U(X_0,Y))], \tag{3.21}$$

where $L_x = X_1 - X_0$, $L_y = Y_1 - Y_0$, $V = L_xL_y$, and we have used the fact that

$$n = \frac{1}{V}\iint\frac{dXdY}{2\pi\ell^2}\,f(U) \tag{3.22}$$

in the lowest order in ℓ^2. Thus, the contribution from the motion of the relative coordinate is cancelled by the bulk contribution of the motion of the center coordinate and only the surface (or edge) contribution remains.

When we consider the low temperature limit, we have to be very careful about the order of taking limit. In the above calculation, the neglected terms are of the order of $(\ell/a)^2\beta U_0$, where U_0 and a are the typical value of the potential U and the typical length characterizing the space variation of U, respectively. We should first take the limit $\ell^2 \to 0$, and then consider the low temperature limit in each term of ℓ^2-expansion series.

4. Quantized Hall Conductivity

In the previous section we have dealt with a bulk problem, but the final expression for σ_{xy} looks like a surface contribution. If we note the unboundedness of the quantity X, however, the r.h.s. of eq.(3.22) is found to be finite in the thermodynamic limit (L_x, $L_y \to \infty$), i.e., it is a bulk quantity. This situation is very similar to that of the diamagnetism.

Let us take the thermodynamic limit. In this case, we should pay a special attention to the edge values of the potential, $U(X_1,Y)$ and $U(X_0,Y)$. If we consider a percolation problem of the equipotential line for a random potential, there will be only one value of energy for which the equipotential line percolates from infinity to infinity. Although there is no rigorous mathematical proof for this statement, physically it is very plausible. When we increase the energy continuously from the minimum value of the potential, the equipotential line will percolate at a certain energy. Similarly, when we decrease the energy from the maximum value of the potential, the percolation of the equipotential line will occur at an energy. These two energies are expected to coincide, if the potential is really random and the system is infinitely large. Thus it is most natural to replace the edge potentials by the percolation energy discussed above, which is represented by U_∞ in the following. These considerations lead to the following expression for σ_{xy},

$$\sigma_{xy} = -\frac{e^2}{h} f(U_\infty) \simeq -\frac{e^2}{h}\theta(\varepsilon_F - U_\infty) \quad , \tag{4.1}$$

where the second approximate equation is valid at very low temperatures. This expression means that σ_{xy} at low temperatures in the strong field limit shows a step-like behaviour as a function of the Fermi energy. This jump will occur at a certain energy near the center of the subband.

If the system size is finite, the edge potential will fluctuate along the sample edges $X = X_0$ and $X = X_1$. Then we have to average over this fluctuation. Assuming the distribution of the values of $U(X_1,Y)$ and $U(X_0,Y)$ is independent of the position, we obtain

$$\sigma_{xy} = -\frac{e^2}{h}\int dU\ f(U)P(U) \quad , \tag{4.2}$$

where $P(U)$ is the distribution function of the edge potential. This gives a gradual increase of the Hall conductivity to the quantized value. In any case, the Hall conductivity takes the quantized value, $-e^2/h$, if the lowest subband is completely filled. This result is consistent with that obtained by Laughlin [15] and Halperin [16] through very general consideration of the gauge invariance. Their arguments, however, do not predict the ε_F-dependence of σ_{xy}.

5. Concluding Remarks

We have shown that the Wigner representation can be powerful method in calculating the transport coefficients in the strong magnetic field limit. In the previous section we derived a step-like behaviour of σ_{xy} as a function of the Fermi energy ε_F. Only the case where ε_F lies within the lowest Landau subband has been considered for simplicity. It is not difficult, however, to extend the present formualtion to include the case where ε_F lies in higher subbands as far as the inter-subband effects are negligible. As a result, the energy dependences of the density of states, $D(\varepsilon)$, and the Hall conductivity, σ_{xy}, are expected to behave as in Fig.1 schematically. In the figure the values of U_1 and U_2 are approximately $U_1 \simeq U_\infty + \hbar\omega_c$ and $U_2 \simeq U_\infty + 2\hbar\omega_c$.

Within the same approximation as in the previous section, it is easy to obtain the expression for σ_{xx}. The result is

$$\begin{aligned}\sigma_{xx} &= \frac{e^2}{V}\frac{\ell^2}{\hbar}\iint\frac{dXdY}{2\pi\ell^2}\frac{\partial f}{\partial Y}X \\ &= \frac{e^2}{h}\frac{1}{L_xL_y}\int_{X_0}^{X_1} dX\ X[f(U(X,Y_1)) - f(U(X,Y_0))] \quad .\end{aligned} \tag{5.1}$$

This vanishes if the edge potentials $U(X,Y_1)$ and $U(X,Y_0)$ can be replaced by U_∞. Even if the edge potentials are allowed to be random variables, σ_{xx} in eq.(5.1) becomes zero after averaging. In order to have a finite value for σ_{xx}, we have to calculate the next order contribution in ℓ^2, which is left for future investigations.

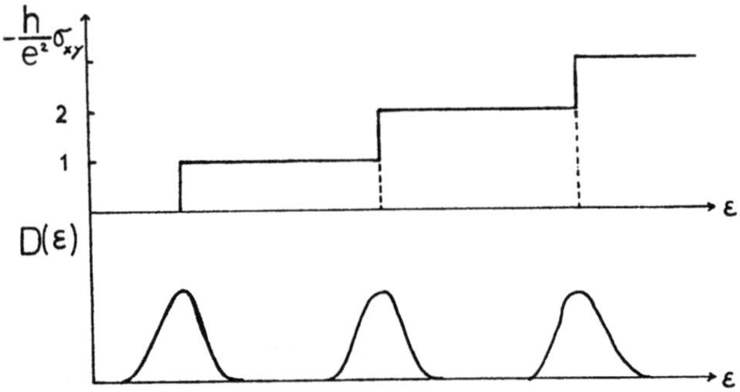

Fig.1 Schematic energy dependences of the density of states, $D(\varepsilon)$, and the Hall conductivity, σ_{xy}, in the strong field limit and at low temperatures

Recently Laughlin discussed the quantized Hall conductivity by employing the gauge invariance argument, and Halperin gave a simpler explanation of Laughlin's argument. In their discussions, the sample form is assumed to be a loop or an annulus. In these configurations, X and Y are not independent variables at the edges. Therefore the present Wigner representation will not be directly applicable to those cases. In order to discuss the loop or annulus configuration, we have to find a more appropriate pair of conjugate variables than X and Y.

Acknowledgements

I would like to thank Professor H. Fukuyama and Dr. D. Yoshioka for fruitful discussions and continuous communications. Many thanks are due to Professor R. Kubo for his critical comments on and deep interest in the present work. In fact, the possibility of applying the Wigner representation to the calculation of the conductivity in the high magnetic field limit was first pointed out by him.

References

[1] E. Abrahams, P.W. Anderson, D.C. Licciardello and T.V. Ramakrishnan: Phys. Rev. Lett. 42 (1979) 673
[2] D.J. Thouless: Phys. Rep. 13C (1974) 93
[3] D. Vollhardt and P. Wölfle: Phys. Rev. B22 (1980) 4666

[4] Y. Ono, D. Yoshioka and H. Fukuyama: J. Phys. Soc. Jpn. 50 (1981) 2143
[5] D. Yoshioka, Y. Ono and H. Fukuyama: J. Phys. Soc. Jpn. 50 (1981) 3419
[6] T. Ando and Y. Uemura: J. Phys. Soc. Jpn. 36 (1974) 959
[7] T. Ando: J. Phys. Soc. Jpn. 36 (1974) 1521;37 (1974) 622
[8] T. Ando, Y. Matsumoto and Y. Uemura: J. Phys. Soc. Jpn. 39 (1975) 279
[9] R. Kubo, S.J. Miyake and N. Hashitsume: *Solid State Physics*, ed. F. Seitz and D. Turnbull (Academic Press, New York,1965) vol.17, p.269
[10] E. Wigner: Phys. Rev. 40 (1932) 749
[11] J.E. Moyal: Proc. Camb. Phil. Soc. 45 (1949) 99
[12] R. Kubo: J. Phys. Soc. Jpn. 19 (1964) 2127
[13] K. Kitahara and Y. Ono: Progr. theor. Phys. 49 (1973) 89
[14] M. Tsukada: J. Phys. Soc. Jpn. 41 (1976) 1466
[15] R.B. Laughlin: Phys. Rev. B23 (1981) 5632
[16] B.I. Halperin: preprint

Part III

Spin Glass

Breaking of Ergodicity in Disordered Systems: The Case of Spin Glasses

G. Toulouse

Laboratoire de Physique de l'Ecole Normale Supérieure
24 rue Lhomond, F-75231 Paris, France

Abstract

The mean field theory of spin glasses has led to novelties, such as the concept of an order function and the violation of linear response theory. An interpretation in terms of breaking of ergodicity is presented, showing equilibrium and remanent properties as intimately connected.

1. Introduction

The mean field theory of spin glasses [1] will be defined for us by the infinite range SK model [2]. In the case of Ising spins, the best existing solution is Parisi's [3], which is expressed in terms of an "order function" q(x), instead of a standard order parameter.

For the purpose of our argument, we shall begin with a reminder of some noteworthy features of the spin glass problem.

The phase diagram (H magnetic field, T temperature) for the Ising SK model is shown in Fig. 1. The transition line corresponds to the onset of spontaneous replica symmetry breaking [1,3]. The existence of a transition line separating a paramagnetic phase from a spin glass phase in real spin glass materials has received experimental support from various groups (Orsay, UCLA, Grenoble, IBM, etc.). It is found experimentally that penetration into the spin glass phase is very different whether one proceeds horizontally or vertically.

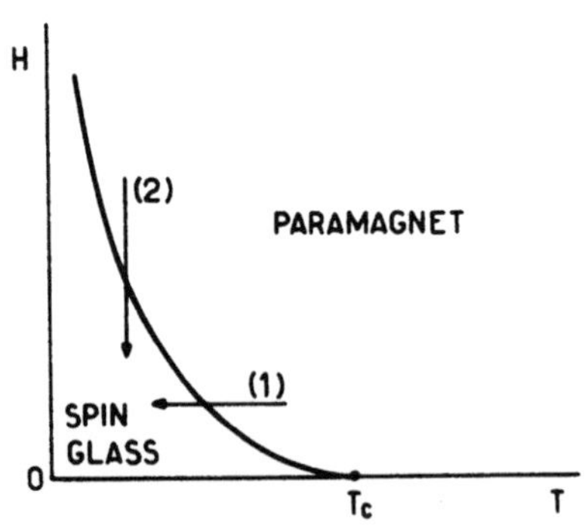

Fig. 1 Phase diagram (H magnetic field, T temperature) for an Ising spin glass in mean field theory. Arrow (1) shows a way of entering the spin glass phase via field-cooling process. Arrow (2) shows another way via isothermal decrease of the field.

Horizontally, i.e. by field-cooling, one is able to observe time-independent properties, that is presumably to remain at equilibrium.

Vertically, i.e. by isothermal decrease of the field, one gets into nonequilibrium states, leading to conspicuous remanence effects.

In Parisi's solution for the Ising spin glass phase, there appears an order function q(x) defined on the interval |0,1|. By a variational calculation, it is found that q(x) varies typically as illustrated in Fig. 2.

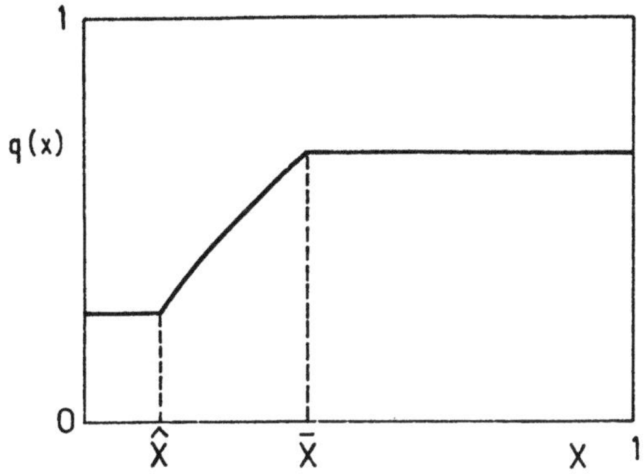

Fig. 2 Typical shape of the order function q(x), presenting two plateaux, in the spin glass phase.

Up to present precision, it is found that q(x) has two plateaux [4]. On the lower plateau

$$q(x;\, T,\, H) \simeq \hat{q}(H) \qquad ,\ 0 < x < \hat{x}; \tag{1}$$

on the higher plateau,

$$q(x;\, T,\, H) \simeq q(T) \qquad ,\ \bar{x} < x < 1. \tag{2}$$

The Edwards-Anderson order parameter

$$q = \overline{\langle Si \rangle^2} \quad , \tag{3}$$

where the bracket means thermal averaging, and the bar disorder averaging, is given by the height of the second plateau

$$q = \underset{x}{\mathrm{Max}}\ q(x) \ . \tag{4}$$

In the paramagnetic phase, the function q(x) is a constant (the two plateaux merge when one reaches the transition line from within the spin glass phase).

Let us now show how a physical quantity such as the magnetic susceptibility is expressed in terms of the order function q(x). The zero-field equilibrium magnetic susceptibility is given by

$$\chi_{eq} = \frac{1 - \int_0^1 q(x)dx}{T} \quad , \tag{5}$$

But this curve $f_m(M)$ also gives the remanent magnetisation (in technical terminology, this is the thermoremanent magnetisation) obtained by isothermally lowering the magnetic field to zero value.

This leads to a relation between the function q(x) and the equilibrium properties, on one side, and the metastable remanent properties, on the other side. More specifically, from formulae (5) and (6), one obtains

$$\chi_{eq}(H = 0) - \chi_{LR}(H = 0) = \frac{1}{T}\int_0^{\bar{x}} [q - q(x)]dx \quad . \tag{11}$$

Up to the factor $\frac{1}{T}$, the r.h.s. of (11) is the hatched area of Fig. 5.

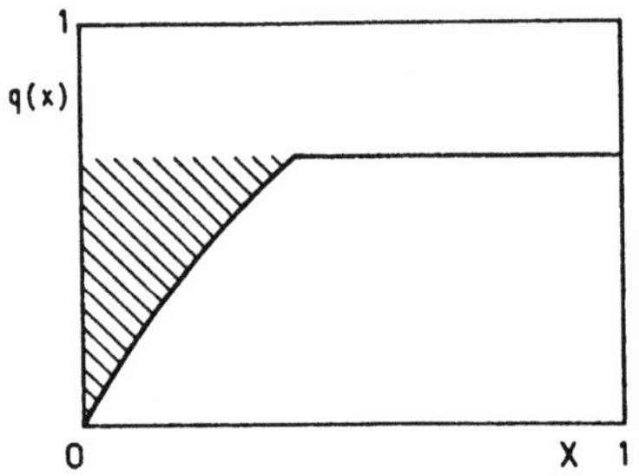

Fig. 5 The hatched area gives the difference between the (zero-field) equilibrium and linear response susceptibilities (up to a factor T).

On the other hand, assuming (this is the simplest approximation one can make) that the valleys have similar shape, one gets

$$\chi_{eq}(H) - \chi_{LR}(H) = \left(\frac{d^2 f_m}{dM^2}\right)^{-1} = \frac{dM_r}{dH} \quad , \tag{12}$$

where $M_r(H)$ is the remanent magnetisation, obtained by isothermal decrease of the field to zero value, starting from equilibrium under field H.

Though this is not yet proved, we suggest that in finite field H, formula (11) may be generalized into

$$\chi_{eq}(H) - \chi_{LR}(H) = \frac{1}{T}\int_{\hat{q}}^{q} x dq \quad , \tag{13}$$

the r.h.s. being equal, up to a factor $\frac{1}{T}$, to the hatched area in Fig. 6.

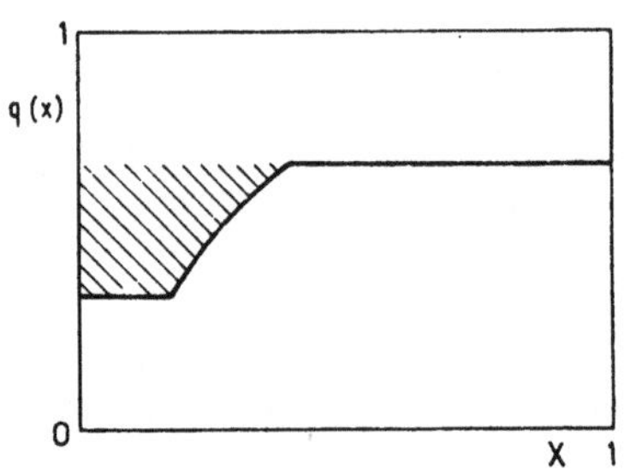

Fig. 6 Generalization of Fig. 5 to the finite field case.

Then, comparison of formulae (12) and (13) shows how the shape of the order function q(x) is directly connected to the remanent properties and to the difference between the equilibrium and linear response susceptibilities.

Using all the information which is presently available, from analytic and numerical sources, on the function q(x), in our problem at hand, it is possible to make detailed predictions concerning the remanent properties, which can be compared with experiments. However, we only wish here to put emphasis on the essentials and we will not discuss these particular developments. Obviously, the picture which is emerging here is likely to be of interest, not only for spin glasses, but for a large class of disordered systems.

4. Discussion

In the aforementioned picture, adequately summarized in Fig. 4, the difference between field cooling and isothermal decrease of the field shows up clearly. Isothermal decrease of the field leads to a metastable state because, at $H_c(T)$, one enters into a particular valley, which is no longer at equilibrium for $H < H_c(T)$. Whereas, in a field-cooling process, lowering of the temperature essentially leads to a widening of the interval $(- M_c(T), + M_c(T))$, and one has at least a chance that the same valley remains tangent to the equilibrium envelope, under constant value of H.

The picture of Fig. 4 appears then as the simplest way of interpreting both the existence of two susceptibilities and the experimental data concerning the two distinct ways of entering the spin glass phase.

The simplicity of this picture stems from the thermodynamic limit (which allows strict ergodicity breaking) and from the mean field character (which allows to treat, in one step, both the equilibrium and the metastable states). It appears as a rather natural generalization of the familiar Van der Waals theory for fluids (this theory is still of interest nowadays, even though it does not include fluctuation and thermal activation effects; the same is expected for the mean field theory of spin glasses).

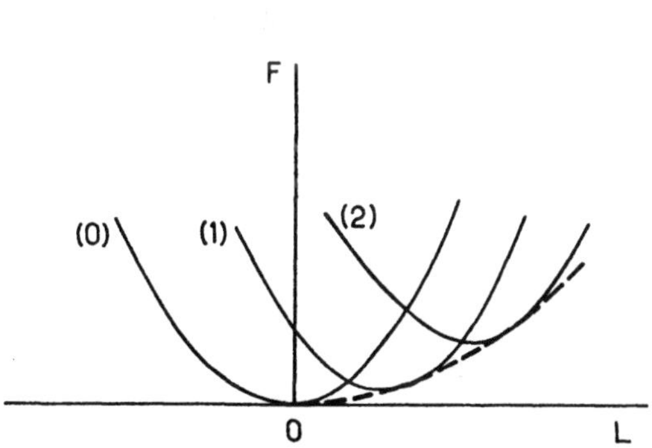

Fig. 7 Schematic plots of the free energy versus angular momentum, for a rotating bucket of superfluid helium. Each valley corresponds to an integer number of vortices. The equilibrium curve is made of piecewise arcs, with a succession of first-order transitions when the angular velocity is varied.

It is actually possible to present a physical system, which is a sort of intermediate link between the simplest case with two valleys and the spin glass case with a continuous distribution of valleys.

Consider a rotating bucket of superfluid helium and plot the free energy as a function of angular momentum. There will be one central valley corresponding to zero vortex number and a denumerable set of valleys corresponding to integer numbers of vortices. In this case, each valley is characterized by a conserved topological number. When the rotation velocity, $\omega = \frac{\partial F}{\partial L}$, is varied, there is a succession of small first-order transitions.

The case of spin glasses, with a continuous distribution of valleys, may then be seen as the limit of a large number of small first-order transitions. Identification of a conserved number, which would characterize each valley (for Ising or for Heisenberg spins), constitutes a desirable goal which has not yet been implemented.

Acknowledgements

I wish to acknowledge the many discussions which I had, at several stages in the past, on this and related matters, with B. Derrida, G. Parisi, J. Vannimenus, J. Souletie. The picture presented above reached this particular form in my mind during the month of July 1981, and I think it is somewhat original, though many of its elements have slowly emerged during the past five years, out of the contributions of very many people, too numerous to be quoted here [1]. It was a pleasure to talk with A. Bray, M. Moore, D. Sherrington, P. Young in Paris and with P. W. Anderson, R. Maynard in Grenoble during the month of September 1981.

References

1 For a recent review on spin glasses, see G. Toulouse, Proceedings of the French Physical Society Conference in Clermont-Ferrand, 1981, to be published in J. Physique.
2 D. Sherrington, S. Kirkpatrick, Phys. Rev. Letters 32, 1792 (1975)
3 G. Parisi, Phys. Reports 67, 25 (1980)
4 J. Vannimenus, G. Toulouse, G. Parisi, J. Physique 42, 565 (1981)

Index of Contributors

Springer Series in

Synergetics

Series Editor: **H. Haken**

Volume 1: **H. Haken**
Synergetics
An Introduction. Nonequilibrium Phase Transitions and Self-Organization in Physics, Chemistry and Biology. 2nd enlarged edition. 1978.
152 figures, 4 tables. XII, 355 pages.
ISBN 3-540-08866-0

Volume 2:
Synergetics
A Workshop. Proceedings of the International Workshop on Synergetics at Schloß Elmau, Bavaria, May 2–7, 1977
Editor: **H. Haken**
1977. 136 figures. VIII, 274 pages.
ISBN 3-540-08483-5

Volume 3:
Synergetics
Far from Equilibrium
Proceedings of the Conference Far from Equilibrium: Instabilities and Structures, Bordeaux, France, September 27–29, 1978
Editors: **A. Pacault, C. Vidal**
1979. 109 figures, 3 tables. IX, 175 pages.
ISBN 3-540-09304-4

Volume 4:
Structural Stability in Physics
Proceedings of Two International Symposia on Applications of Catastrophe Theory and Topological Concepts in Physics, Tübingen, Federal Republic of Germany, May 2–6 and December 11–14, 1978
Editors: **W. Güttinger, H. Eikemeier**
1979. 108 figures, 8 tables. VIII, 311 pages.
ISBN 3-540-09463-6

Volume 5:
Pattern Formation by Dynamic Systems and Pattern Recognition
Proceedings of the International Symposium on Synergetics at Schloß Elmau, Bavaria, April 30–May 5, 1979
Editor: **H. Haken**
1979. 156 figures, 16 tables. VIII, 305 pages.
ISBN 3-540-09770-8

Volume 6:
Dynamics of Synergetic Systems
Proceedings of the International Symposium on Synergetics, Bielefeld, Federal Republic of Germany, September 24–29, 1979
Editor: **H. Haken**
1980. 146 figures, some in color, 5 tables.
VIII, 271 pages. ISBN 3-540-09918-2

Volume 7: **L. A. Blumenfeld**
Problems of Biological Physics
1981. 38 figures. IX, 224 pages. ISBN 3-540-10401-1

Volume 8:
Stochastic Nonlinear Systems
in Physics, Chemistry, and Biology
Proceedings of the Workshop, Bielefeld, Federal Republic of Germany, October 5–11, 1980
Editors: **L. Arnold, R. Lefever**
1981. 48 figures. VIII, 237 pages.
ISBN 3-540-10713-4

Volume 9:
Numerical Methods in the Study of Critical Phenomena
Proceedings of a Colloquium, Carry-le-Rouet, France, June 2–4, 1980
Editors: **J. Della Dora, J. Demongeot, B. Lacolle**
1981. 83 figures. IX, 269 pages. ISBN 3-540-11009-7

Volume 10: **Yu. L. Klimontovich**
The Kinetic Theory of Electromagnetic Processes
ISBN 3-540-11458-0. In preparation

Volume 11:
Chaos and Order in Nature
Proceedings of the International Symposium on Synergetics at Schloß Elmau, Bavaria, April 27–May 2, 1981
Editor: **H. Haken**
1981. 134 figures. VIII, 275 pages.
ISBN 3-540-11101-8

Volume 12:
Nonlinear Phenomena in Chemical Dynamics
Proceedings of an International Conference, Bordeaux, France, September 7–11, 1981
Editors: **C. Vidal, A. Pacault**
1981. 124 figures. X, 280 pages. ISBN 3-540-11294-4

Volume 13: **C. W. Gardiner**
Handbook of Stochastic Methods
for Physics, Chemistry and the Natural Sciences
1982. Approx. 29 figures. Approx. 480 pages.
ISBN 3-540-11357-6. In preparation

Volume 14: **W. Weidlich, G. Haag**
Dynamics of Interacting Populations in Society
ISBN 3-540-11358-4. In preparation

Volume 15: **W. Horsthemke, R. Lefever**
Nonequilibrium Transitions Induced by External Noise
ISBN 3-540-11359-2. In preparation

Volume 16: **L. A. Blumenfeld**
Physics of Bioenergetic Processes
ISBN 3-540-11417-3. In preparation

Springer-Verlag
Berlin
Heidelberg
New York

MIX
Papier aus verantwortungsvollen Quellen
Paper from responsible sources
FSC® C105338

If you have any concerns about our products,
you can contact us on
ProductSafety@springernature.com

In case Publisher is established outside the EU,
the EU authorized representative is:
Springer Nature Customer Service Center GmbH
Europaplatz 3, 69115 Heidelberg, Germany

Printed by Libri Plureos GmbH
in Hamburg, Germany